MODELLING COMMAND AND CONTROL

Modelling Command and Control
Event Analysis of Systemic Teamwork

NEVILLE A. STANTON
CHRIS BABER
DON HARRIS
Human Factors Integration Defence Technology Centre

ASHGATE

© Neville A. Stanton, Chris Baber and Don Harris 2008

All rights reserved. No part of this publication may be reproduced, stored in a retrieval system or transmitted in any form or by any means, electronic, mechanical, photocopying, recording or otherwise without the prior permission of the publisher.

Neville A. Stanton, Chris Baber and Don Harris have asserted their moral right under the Copyright, Designs and Patents Act, 1988, to be identified as the authors of this work.

Published by
Ashgate Publishing Limited
Gower House
Croft Road
Aldershot
Hampshire GU11 3HR
England

Ashgate Publishing Company
Suite 420
101 Cherry Street
Burlington, VT 05401-4405
USA

Ashgate website: http://www.ashgate.com

British Library Cataloguing in Publication Data
Stanton, Neville, 1960-
   Modelling command and control : event analysis of systematic teamwork. - (Human factors in defence)
   1. Command and control systems - Simulation methods
   2. Command of troops - Simulation methods
   I. Title II. Baber, Christopher, 1964- III. Harris, Don, 1961-
   355.3'3041

Library of Congress Cataloging-in-Publication Data
Stanton, Neville, 1960-
   Modelling command and control : event analysis of systemic teamwork / by Neville A. Stanton, Chris Baber and Don Harris.
       p. cm. -- (Human factors in defence)
   Includes bibliographical references and index.
   ISBN 978-0-7546-7027-8
1. Command and control systems. 2. Command and control systems--Data processing. I. Baber, Christopher, 1964- II. Harris, Don, 1961- III. Title.

UB212.S72 2008
355.3'3041011--dc22

2007030959

ISBN 978-07546-7027-8

Printed and bound in Great Britain by
MPG Books Ltd, Bodmin, Cornwall.

Contents

List of Figures	*vii*
List of Tables	*xi*
Preface	*xv*
Acknowledgements	*xix*
Senior Author Biographies	*xxi*
Contributing Authors	*xxiii*

Chapter 1	**Overview**		**1**
	Chapter 2	Modelling Command and Control	1
	Chapter 3	Event Analysis of Systemic Teamwork	2
	Chapter 4	Case Study at HMS Dryad	3
	Chapter 5	Case Study in RAF Boeing E3D Sentry	4
	Chapter 6	Case Study in Battle-Group HQ	4
	Chapter 7	Development of a Generic Process Model of Command and Control	5

Chapter 2	**Modelling Command and Control**	**7**
	Purpose and Scope	7
	Models	7
	Specification	10
	Structural Models of Command and Control	13
	Network Models	19
	Dynamic Models	25
	Agent Models	32
	Socio-technical Models	36
	Example: Contextual Control Model	43
	Summary of Socio-technical Models	45
	Summary of Modelling Review	45

Chapter 3	**Event Analysis of Systemic Team-work**	**49**
	EAST Review	49
	Methods Review	51
	Summary	117

Chapter 4	**Case Study at HMS Dryad**	**119**
	Introduction	121
	Observations	124
	Conclusions	154

Chapter 5	**Case Study in RAF Boeing E3D Sentry**	**157**
	Introduction	157
	Observations	159
	Propositional Networks	172
	Conclusions	177
Chapter 6	**Case Study in Battle Group HQ**	**181**
	Introduction	181
	Findings	192
	Conclusions	219
Chapter 7	**Development of a Generic Process Model of Command and Control**	**221**
	Three Domains for Command and Control	221
	Common Features of the Domains and Application of Command and Control	225
	Taxonomies of Command and Control Activities	225
	Construction of the Model	231
	Network Enabled Capability	233
	Model Validity	234
	Summary	235
	Conclusions	235

Bibliography   *239*
Index   *245*

List of Figures

Figure 2.1	SAS-050 Space of C^2	12
Figure 2.2	An adapted version of Lawson's model of the command and control process	15
Figure 2.3	The OODA loop	16
Figure 2.4	Example of closed loop feedback control system	17
Figure 2.5	Network Centric Operations Conceptual Framework	18
Figure 2.6	The N-squared (N^2) chart from Hitchens (2000): A simplistic generic model of command and control	20
Figure 2.7	Pyramid structure metrics	21
Figure 2.8	Prototypical social networks	22
Figure 2.9	Propositional network from the air traffic control work domain	24
Figure 2.10	Centralised architecture without information sharing	27
Figure 2.11	Split architecture without information sharing	27
Figure 2.12	Distributed architecture without information sharing	29
Figure 2.13	Negotiated architecture without information sharing	29
Figure 2.14	Distributed with information sharing	31
Figure 2.15	Performance of different network structures	31
Figure 2.16	Architecture of a semiotic agent	35
Figure 2.17	Simulated agents produce emergent flocking behaviour within a computer simulation	35
Figure 2.18	Kotter's model of organisational dynamics (1978)	37
Figure 2.19	Smalley's functional command and control model	38
Figure 2.20	Definition of the abstraction decomposition space (Rasmussen, 1986)	40
Figure 2.21	Rasmussen's decision-ladder applied to a command and control domain	41
Figure 2.22	Example output of strategies analysis. Shaded regions refer to a specific role or actor; the numbered circles refer to task steps	42
Figure 2.23	Hollnagel's contextual control model	44
Figure 2.24	NEC benefits chain	47
Figure 3.1	Social network diagram	60
Figure 3.2	CUD extract	61
Figure 3.3	OSD glossary	61
Figure 3.4	OSD extract	62
Figure 3.5	Propositional network for objects referred to in CDM tables	68
Figure 3.6	Propositional network for CDM phase one	68
Figure 3.7	Propositional network for CDM phase two	69

Figure 3.8	Propositional network for CDM phase three	69
Figure 3.9	Propositional network for CDM phase four	70
Figure 3.10	Hazardous chemical spillage event flowchart	76
Figure 3.11	CUD template	95
Figure 3.12	Comms usage diagram for an energy distribution task	95
Figure 3.13	Return to service social network diagram	98
Figure 3.14	Example OSD template	104
Figure 3.15	Example propositional network	114
Figure 4.1	Illustration of workstations onboard a Type-23 frigate	121
Figure 4.2	Seating lay out of Type-23 frigate operations room	122
Figure 4.3	Combined task model for the air, subsurface and surface threat	125
Figure 4.4	Social network encompassing all three scenarios	132
Figure 4.5	OSD air threat scenario	133
Figure 4.6	OSD subsurface threat scenario	135
Figure 4.7	OSD surface threat mission	137
Figure 4.8	Propositional network for the air threat task	139
Figure 4.9	Plan resources and strategy	140
Figure 4.10	Identify and classify targets	140
Figure 4.11	Assess threat and locate targets	141
Figure 4.12	Control external resources	141
Figure 4.13	Engage targets	142
Figure 4.14	Posture platform and attack	142
Figure 4.15	Re-allocate assets and weapons to new targets	143
Figure 4.16	Shared knowledge objects of the air threat scenario	143
Figure 4.17	Propositional network for the sub-surface threat task	144
Figure 4.18	Plan resources and strategy	144
Figure 4.19	Identify and classify targets	145
Figure 4.20	Assess threat and locate targets	145
Figure 4.21	Control external resources	146
Figure 4.22	Engage targets	146
Figure 4.23	Posture platform and attack	147
Figure 4.24	Re-allocate assets and weapons to new targets	147
Figure 4.25	Shared knowledge objects of the air threat scenario	148
Figure 4.26	Propositional network for the surface threat task	148
Figure 4.27	Plan resources and strategy	149
Figure 4.28	Identify and classify targets	149
Figure 4.29	Assess threat and locate targets	150
Figure 4.30	Control external resources	150
Figure 4.31	Engage targets	151
Figure 4.32	Posture platform and attack	151
Figure 4.33	Re-allocate assets and weapons to new targets	152
Figure 4.34	Shared knowledge objects of the air threat scenario	152
Figure 5.1	Typical seating plan onboard an RAF E3D	158

List of Figures

Figure 5.2	Mission crew structure	159
Figure 5.3	Illustration of workstations on board the E3D	161
Figure 5.4	Task model of E3D mission	161
Figure 5.5	Social network diagram constructed from the standard operating procedures	163
Figure 5.6	FA (1) social network diagram	168
Figure 5.7	FA (2) social network diagram	168
Figure 5.8	OCA social network diagram	169
Figure 5.9	SO4 social network diagram	169
Figure 5.10	SC social network diagram	170
Figure 5.11	FA(1) operational sequence diagram	171
Figure 5.12	Propositional network for the E3D mission	172
Figure 5.13	Assume station (TD)	173
Figure 5.14	Manage airborne battle (FA, WC, TD)	173
Figure 5.15	Coordinate crew	174
Figure 5.16	Manage communications (SO, LM, TD, CO, CT, SC, RT)	174
Figure 5.17	Ensure operational procedures are followed (TD)	175
Figure 5.18	Manage self defence (TD)	175
Figure 5.19	Manage operations/comms security (TD)	176
Figure 5.20	Control surveillance/operate ESM equipment (SC, ESM)	176
Figure 5.21	Overlap of knowledge objects	177
Figure 6.1	Simplified command hierarchy	181
Figure 6.2	IPB system and products	183
Figure 6.3	Illustration of simplified BAE overlay showing water features and restricted terrain	184
Figure 6.4	Illustration of BAE overlay showing possible avenues of approach to engage enemy force	184
Figure 6.5	Illustration of a situation overlay; icons show formation of enemy forces, arrows show possible routes/courses of action, dotted lines indicate time phases from the enemy's current position	185
Figure 6.6	Illustration of the commander's effects schematic; the battle winning idea expressed in terms of effects on the enemy	186
Figure 6.7	Illustration of a draft decision support overlay showing some key features of the BAE in relation to target and named areas of interest (TAI and NAI respectively) and decision points (stars)	187
Figure 6.8	Illustration of a completed synchronisation matrix showing the time line (time based decision points) upon which enemy and own courses of action are mapped along with the specific activities of the various force elements	189
Figure 6.9	Example of operational graphics	190
Figure 6.10	Relationship between the combat estimate and Wargaming	191
Figure 6.11	Task network of observed military command and control activities derived from HTA	193
Figure 6.12	Photograph of the command tent	197

Figure 6.13	Social network diagram illustrating a systems level view of the combat estimate scenario	200
Figure 6.14	'Briefing': illustration of social activity and communication. The one way flow of information from commanding officer to subordinate staff is evident as is the close coupling between nodes	201
Figure 6.15	'Reviewing': illustration of social activity and communication. The collaborative, 2-way nature of communication is evident as are the ad-hoc 'open' links to planning materials (shown by dotted links)	201
Figure 6.16	'Semi-Autonomous working': illustration of social activity and communication. The agents are primarily linked to the planning materials that are being worked upon with 'ad-hoc' open channels to other Headquarters staff (shown by the dotted line)	202
Figure 6.17	Illustration of social network where the communication links between nodes have been annotated with the media or modality that facilitates it. The sub-units and higher formation are geographically remote from the Battlegroup Headquarters	204
Figure 6.18	Key to enhanced OSD symbology	206
Figure 6.19	OSD representation of the 'prepare to make a plan' phase	207
Figure 6.20	OSD representation of the phase dealing with question 1 of the Combat Estimate	207
Figure 6.21	OSD representation of the phase dealing with questions 2 and 3 of the Combat Estimate	208
Figure 6.22	OSD representation of the phase dealing with questions 4 and 5 of the Combat Estimate	208
Figure 6.23	OSD representation of the phase dealing with questions 6 and 7 of the Combat Estimate	209
Figure 6.24	OSD representation of the concluding phases of the Combat Estimate and execution of the plan	210
Figure 6.25	Overview of systems level knowledge for the CAST scenario represented via propositional network	211
Figure 6.26	Knowledge objects associated with question 1 of the Combat Estimate technique	212
Figure 6.27	Knowledge objects associated with questions 2 and 3 of the Combat Estimate technique	213
Figure 6.28	Knowledge objects associated with question 4 of the Combat Estimate technique	214
Figure 6.29	Knowledge objects associated with question 5 of the Combat Estimate technique	215
Figure 6.30	Knowledge objects associated with questions 6 and 7 of the Combat Estimate technique	216
Figure 6.31	Knowledge object activities with putting the plan into effect	217
Figure 7.1	Generic process model of command and control	232

List of Tables

Table 2.1	Sample of mathematical metrics for key constructs in command and control	14
Table 2.2	Propositional network metrics to detect emergent property of SA in relation to key knowledge objects	23
Table 2.3	Relating SAS-050 variables to Scud Hunt Coefficients	28
Table 2.4	Mapping Dekker's SAS-050 models	30
Table 2.5	Worked example of an abstraction decomposition space from a militaristic work domain	40
Table 2.6	Summary of modelling perspectives and broad emergent properties	46
Table 2.7	Informal classification of model types/typologies with defined NEC benefit criteria	48
Table 3.1	EAST analyses	50
Table 3.2	Summary of EAST methods review	52
Table 3.3	Agents involved in switching scenario	58
Table 3.4	Switching scenario CDA results	58
Table 3.5	Extract of CDA analysis	59
Table 3.6	Agent association matrix	60
Table 3.7	SNA results	60
Table 3.8	Operational loading results	62
Table 3.9	CDM probes	63
Table 3.10	CDM Phase 1: First issue of instructions	64
Table 3.11	CDM Phase 2: Deal with switching requests	65
Table 3.12	CDM Phase 3: Perform Isolation	66
Table 3.13	CDM Phase 4: Report back to NOC	67
Table 3.14	Observation transcript extract	73
Table 3.15	Example HTA plans	81
Table 3.16	CDA teamwork taxonomy	87
Table 3.17	CUD summary table	89
Table 3.18	Agents involved in the return to service scenario	100
Table 3.19	Agent association matrix	100
Table 3.20	Agent centrality (B-L Centrality)	100
Table 3.21	Agent sociometric status	101
Table 4.1	Glossary of abbreviations	120
Table 4.2	The main agents involved in the mission	123
Table 4.3	Air threat scenario CDA results	126

Table 4.4	Air threat scenario CDA results in HTA stages	126
Table 4.5	PWO subsurface scenario CDA results	127
Table 4.6	Subsurface threat scenario CDA results in HTA stages	127
Table 4.7	PWO surface scenario CDA results	129
Table 4.8	Surface threat scenario CDA results in HTA stages	129
Table 4.9	Extract of CDA analysis	130
Table 4.10	Operational loading for the air threat scenario	134
Table 4.11	Operational loading for the subsurface threat scenario	134
Table 4.12	Operational loading for the surface threat scenario	136
Table 4.13	Analysis of core knowledge objects against the seven phases of operation	154
Table 5.1	E3D mission scenario CDA results	162
Table 5.2	E3D mission CDA results in HTA stages	162
Table 5.3	Extract of CDA analysis where 1 = low, 2 = medium and 3 = high	164
Table 5.4	List of agents involved in the E3D operations	166
Table 5.5	Matrix showing association between agents on the E3D	166
Table 5.6	Comparison for positional centrality (degree dimension)	167
Table 6.1	Illustration of a partially completed DSO matrix. The matrix lists the order in which TAIs are to be dealt with and the associated resources to be used	188
Table 6.2	Illustration of a simplified decision support matrix	189
Table 6.3	Coordination demand dimensions	194
Table 6.4	CAST scenario CDA results	195
Table 6.5	CAST scenario CDA results in HTA stages	196
Table 6.6	Advantages and disadvantages of existing comms media	198
Table 6.7	Network metrics illustrating centrality (key agents in the scenario) and density (network connectivity) for the social network as a whole	203
Table 6.8	Network metrics illustrating centrality (key agents in the scenario) and density (network connectivity) for the activity stereotypes of Briefing, Reviewing and Semi-Autonomous Working	203
Table 6.9	Technology/facilitation/modality matrix. Shading shows the match between technology and modality	205
Table 6.10	Summary of key knowledge objects active within each scenario	218
Table 7.1	Taxonomy of command and control activities	226
Table 7.2	The 'receive' activities taxonomy	227
Table 7.3	The 'planning' activities taxonomy	228
Table 7.4	The 'rehearsal' activities taxonomy	229
Table 7.5	The 'communicate' activities taxonomy	229
Table 7.6	The 'request' activities taxonomy	230

Table 7.7	The 'monitor' activities taxonomy	230
Table 7.8	The 'review' activities taxonomy (The taxonomies are related to comprehensive HTAs of each individual scenario in question. These can be found separately in the individual EAST analysis reports that deal with each live scenario.)	231

Preface

Readers of this book might be interested to know something of the genesis of ideas and projects that led to its conception. The basic idea to develop a new approach to the analysis and representation of command and control began early in 2002, when we decided to form a consortium to bid for the Ministry of Defence research contract for a Human Factors Integration Defence Technology Centre (HFI DTC). At that time there was a paradigm shift in defence research spending, with the desire to set up consortia based on collaboration between academia and industry. We formed a consortium led by Aerosystems International, with Birmingham, Brunel and Cranfield Universities together with Lockheed Martin, MBDA and SEA. In the competitive tendering process, we had to develop a set of research ideas that would define the work of the consortium for three years, from 2003 to 2006. One of the strands of work focused on Command and Control. Part of this work was to be focused on the analysis of military Command and Control using a new approach (which was to be developed). Because part of the remit of the HFI DTC was to examine spin-in to the defence domain from the civil sector (and vice versa), we looked at Command and Control in the emergency services and other civil applications. However, this book focuses almost exclusively on research in the military domain.

As history testifies, we were successful in winning the initial HFI DTC contract and have also won a follow on contract, which takes our programme of research from April 2003 to March 2009. From those humble beginnings in 2002, we have developed a new approach to the description and analysis of Command and Control called Event Analysis of Systemic Team-work (EAST) as well as a sister method called Workload, Error, Situation awareness, Time and Team-work (WESTT). This book presents the EAST method and applications of the method in the Navy, Air Force and Army Command and Control. EAST brings together a collection of Human Factors methods in an integrated manner, rather than developing completely new methods from scratch. The novelty of the approach is in the way in which the various methods have been integrated and the network models that the application of the methods produce as outputs. These models enable the analyst to consider the Command and Control system under scrutiny in terms of the task, social and knowledge networks, as well as the inter-relations between these networks.

At the time of writing the HFI DTC comprises:

Aerosystems International

Dr David Morris
Dr Karen Lane
Stephen Brackley
Linda Wells
Kevin Bessell
Nic Gibbs
Kelly Maddock-Davies

The University of Birmingham

Dr Chris Baber
Professor Bob Stone
Dr Huw Gibson
Dr Rob Houghton
Richard McMaster
Dr James Cross
Robert Guest

Brunel University

Professor Neville A. Stanton
Paul S. Salmon
Dr Guy H. Walker
Dr Dan Jenkins
Amardeep Ajula
Kirsten Revell

Cranfield University

Dr Don Harris
Dr John Huddlestone
Dr Geoff Hone
Jacob Mulenga
Ian Whitworth
Andy Farmilo
Antoinette Caird-Daley
Jon Pike
Louise Forbes

Lockheed Martin UK

Mick Fuchs
Lucy Mitchell
Mark Linsell
Ben Leonard
Rebecca Stewart

Systems Engineering and Assessment Ltd

Pam Newman
Dr Anne Bruseberg
Dr Iya Solodilova-Whiteley
Mel Lowe
Ben Dawson
Jonathan Smalley
Dr Anne Bruseberg

DSTL

Geoff Barrett
Bruce Callander
Jen Clemitson
Colin Corbridge
Roland Edwards
Alan Ellis
Jim Squire
Debbie Webb

MBDA Missile Systems

Dr Carol Mason
Grant Hudson
Roy Dymott

This book may be approached in different ways by different readers. For an overview of the entire book, read chapter one. For those wishing to get to grips with the modelling literature, read chapter two for a review and chapter seven for our own Command and Control model. Those readers that want to apply the EAST method to their domain, we recommend reading chapter three plus one of the case studies in chapters four, five or six. For those readers who want to understand some aspects of Command and Control in the military domain, read chapters four, five and six. We have tried to write the book so that each of the chapters can stand alone, but inevitably there is some inheritance from previous chapters, which may require the reader to dip in and out to make the most of each chapter. The index section should help guide the reader.

The research work and spirit of the HFI DTC is such that collaboration between all of the organisations was integral to our approach, combining the best mix of academic and industrial knowledge and skills. We have learnt a lot about the military and about collaborating in this new approach to research funding in the Ministry of Defence. The willingness to participate and the can-do attitude of our Armed Forces is inspiring. We are very grateful, in particular, to the personnel at HMS Dryad in Portsmouth, CASTT at the Land Warfare Centre in Warminster and the E3D Sentry crews at RAF Waddington (8 and 23 Squadron) who have made this research possible. We hope that you, the reader, will find this book both useful and interesting in your own research endeavours. We have certainly enjoyed all of the experiences that have led to us bringing the fascinating area of Human Factors research onto the printed pages in front of you.

Neville A. Stanton
Chris Baber
Don Harris

Acknowledgements

The Human Factors Integration Defence Technology Centre is a consortium of defence companies and UK Universities working in cooperation on a series of defence related projects. The consortium is led by Aerosystems International and comprises Birmingham University, Brunel University, Cranfield University, Lockheed Martin, MBDA and SEA.

We are grateful to DSTL who have managed the work of the consortium, in particular to Geoff Barrett, Bruce Callander, Jen Clemitson, Colin Corbridge, Roland Edwards, Alan Ellis, Jim Squire and Debbie Webb.

This work from the Human Factors Integration Defence Technology Centre was part-funded by the Human Sciences Domain of the UK Ministry of Defence Scientific Research Programme.

Senior Author Biographies

Professor Neville A. Stanton
HFI DTC, BIT Lab, School of Engineering and Design
Brunel University, Uxbridge, Middlesex
UB8 3PH
UK
neville.stanton@brunel.ac.uk
http://www.brunel.ac.uk/about/acad/sed/sedres/dm/erg/

Professor Stanton holds a Chair in Human-Centred Design and has published over 75 international academic journal papers and 10 books on human-centred design. He was a Visiting Fellow of the Department of Design and Environmental Analysis at Cornell University in 1998. In 1998 he was awarded the Institution of Electrical Engineers Divisional Premium Award for a co-authored paper on Engineering Psychology and System Safety. The Ergonomics Society awarded him the prestigious Otto Edholm medal in 2001 for his contribution to basic and applied ergonomics research. Professor Stanton is on the editorial boards of *Ergonomics*, *Theoretical Issues in Ergonomics Science* and the *International Journal of Human Computer Interaction*. Professor Stanton is a Chartered Occupational Psychologist registered with The British Psychological Society, a Fellow of The Ergonomics Society. He has a BSc in Occupational Psychology from Hull University, an MPhil in Applied Psychology from Aston University, and a PhD in Human Factors, also from Aston.

Dr Chris Baber
HFI DTC, Electronic, Electrical and Computer Engineering
The University of Birmingham, Birmingham
B15 2TT
UK
c.baber@bham.ac.uk
http://www.eee.bham.ac.uk/baberc

Dr Baber is a Reader in Interactive Systems Design, a member of the Human Interface Technology Group, Educational Technology Group and the Pervasive Computing Groups within the School of Electronic and Electrical Engineering, and an Associate Member of the Sensory Motor Neuroscience Group in the School of Psychology. He joined the University in 1990 as a lecturer on the MSc Work Design and Ergonomics course, before moving to his new post in 1999. In 2002, he was promoted to Senior Lecturer, and to Reader in 2004. Dr Baber has been involved in ergonomics research since 1987. His main area of interest is in the ways in which people make sense of and make use of 'everyday technology' and how 'everyday' skills can be brought to bear in people's interaction with computers. At present, he is working on two funded projects: Wearable Computers for Crime Scene Investigation (funded by EPSRC 2004–2007) and the Defence Technology Centre for Human Factors Integration (2003–2009). Work on the latter project covers further development of wearable and mobile computers, design and development of a performance-modelling tool, and research into distributed operations. His early research was into speech-based interaction with computers, and he was particularly interested in issues of error correction and dialogue design.

Dr Don Harris
HFI DTC, Department of Human Factors
School of Engineering, Cranfield University
Cranfield, Bedford
MK43 0AL
UK
d.harris@cranfield.ac.uk
http://www.cranfield.ac.uk/soe/hf/biog_d_harris.htm

Don Harris is a Reader and Course Director for the MSc in Human Factors in Health and Safety at Work and MSc Ergonomics and Safety at Work. He is Director of Flight Deck Design and Aviation Safety Group and of the Defence Human Factors Group. His principal teaching and research interests lie in the design and evaluation of flight deck control and display systems, accident investigation and analysis, system safety and flight simulation and training. Don is a Fellow of both the Ergonomics Society and the Higher Education Academy. He is also a chartered Psychologist. He is Chairman of the International Conference series on Engineering Psychology and Cognitive Ergonomics. He sits on the editorial boards of the *International Journal of Applied Aviation Studies* and *Cognition, Technology and Work* and is Co-Editor in Chief of the Journal *Human Factors and Aerospace Safety* (published by Ashgate). In 2006 Don was a member of a team that received the Royal Aeronautical Society Bronze Award for advances in Aerospace and was invited to address the Chinese Academy of Sciences in Beijing.

Contributing Authors

Dan Jenkins, Paul S. Salmon, Guy H. Walker and Mark Young
HFI DTC, BIT Lab, School of Engineering and Design
Brunel University, Uxbridge, Middlesex, UB8 3PH UK

Rob Houghton and Richard McMaster
HFI DTC, Electronic, Electrical and Computer Engineering
The University of Birmingham, Birmingham, B15 2TT UK

Alison Kay
HFI DTC, Department of Human Factors
School of Engineering, Cranfield University
Cranfield, Bedford, MK43 0AL UK

Linda Wells
HFI DTC, Aerosystems International Ltd
Alvington, Yeovil, Somerset, BA22 8UZ UK

Roy Dymott
HFI DTC, MBDA Missile Systems
Golf Course Lane, Filton, Bristol, BS34 7QW UK

Mark Linsell and Rebecca Stewart
HFI DTC, Lockheed Martin UKIS
Building 7000, Langstone Technology Park
Langstone, Havant, Hampshire, PO9 1SW UK

Geoff Hoyle, Mel Lowe and Kerry Tatlock
SEA House, Building 660, The Gardens
Bristol Business Park, Coldharbour Lane, Bristol, BS16 1EJ UK

Chapter 1

Overview

The Defence Technology Centre for Human Factors Integration (DTC HFI) is a research consortium comprised of academic institutions and defence companies, part funded by the Human Sciences Domain of the UK Ministry of Defence Scientific Research Programme. Human Factors Integration is about:

> ... providing a balanced development of both the technical and human aspects of equipment provision. It provides a process that ensures the application of scientific knowledge about human characteristics through the specification, design and evaluation of systems.' (MoD, 2000, p.6)

The aim of the book is to show the application of the HFI-DTC's Event Analysis for Systemic Teamwork (EAST) methodology to military Command and Control applications. The chapters in the book report on a series of investigations from a Human Factors perspective. The book begins with an overview of Command and Control models. Then the EAST methodology is explained. This is followed by three case studies in different domains: Sea, Air and Land. The final chapter draws the material from the case studies, and other research, to present a generic activities model of Command and Control.

Chapter 2   Modelling Command and Control

With contributions from Rob Houghton, Paul S. Salmon and Guy H. Walker

Since its inception, just after the Second World War, Human Factors research has paid special attention to the issues surrounding human control of systems. Command and Control environments continue to represent a challenging domain for Human Factors research. We take a broad view of Command and Control research, to include C2 (Command and Control), C3 (Command, Control and Communication), and C4 (Command, Control, Communication and Computers) as well as Human Supervisory Control paradigms. This book will present case studies in diverse military applications (for example, land, sea and air) of command and control. While the domains of application are highly diverse, many of the challenges they face share interesting similarities.

Chapter 3 Event Analysis of Systemic Teamwork

With contributions from Dan Jenkins, Paul S. Salmon and Guy H. Walker

The EAST methodology is particularly suited to the analysis of team-based or collaborative activity, such as that seen in C4i environments (command, control, communication, computers and intelligence). The method was originally developed for this purpose and applications so far have proved extremely successful, highlighting its suitability for such applications. The EAST methodology is an exhaustive set of data collection, analysis and representation methods. A number of different analyses are conducted and various perspectives on the scenario(s) under analysis are offered. In its present form, the EAST methodology offers the following analyses of a particular C4i scenario:

- A step-by-step (goals, sub-goals, operations and plans) description of the activity in question.
- A definition of roles within the scenario.
- An analysis of the agent network structure involved (for example network type and density).
- A rating of co-ordination between agents for each team-based task step and an overall co-ordination rating.
- An analysis of the current technology used during communications between agents and also recommendations for novel communications technology.
- A description of the task in terms of the flow of information, communications between agents, the activity conducted by each agent involved and a timeline of activity.
- An analysis of agent centrality, sociometric status and betweenness within the network involved in the scenario.
- A definition of the key agents involved in the scenario.
- A cognitive task analysis of operator decision making during the scenario.
- A definition of the knowledge objects (information, artefacts etc) required and the knowledge objects used during the scenario.
- A definition of shared knowledge or shared situation awareness during the scenario.

The EAST methodology is relatively simple to use. The method requires an initial understanding of HF and experience in the application of HF methods. However, from analyst reports it can be concluded that the EAST methodology is relatively simple to apply. The method's ease of use is heightened when compared to the exhaustive output that is generated from an EAST analysis. The EAST methodology is generic and can be applied in any domain in which collaborative activity takes place. As an overall conclusion, it is felt by the authors that the EAST methodology has been successful in its applications thus far and the methodology is perfectly suited to the analysis of Command and Control activity. Each EAST application has produced valid and useful results that are currently forming the basis for the development of a Command and Control model.

Chapter 4   Case Study at HMS Dryad

With contributions from Roy Dymott, Rob Houghton, Geoff Hoyle, Mark Linsell, Richard McMaster, Paul S. Salmon, Rebecca Stewart, Guy H. Walker and Mark Young

This study was conducted in order to analyse C4i in the Royal Navy domain. Researchers were given access to one of the Navy's training establishments – the Maritime Warfare School – HMS Dryad in Southwick, Hampshire. Observations were made during Command Team Training (CTT). This programme involved training the Command Team of a warship in the skills that would be necessary for them to defend their ship in a multi-threat environment.

The aim of this study was to apply the EAST methods to assess the communication and command on board the Type 23 frigate. The study uses the methodology to explore a complex communication system between sixteen team members where effective communication, decision making and coordination are essential to task success, for this highly distributed communication network. Three scenarios (air threat, subsurface threat and surface threat) were observed and analysed. The scenarios were different, and due to the complexity of the task only individual crew members could only be observed at any time as opposed to the scenario being viewed as a whole. However an overall idea of communications could be ascertained from these individual observations.

Data are presented in the form of Social Network Analysis, Coordination Demand Analysis and Propositional Network Analysis. Shared situation awareness is also considered in this case study using the propositional networks. Knowledge objects are identified for the whole of the mission as well as for the phases of the mission. This gives an indication of where there is sharing of information. The networks also suggest where knowledge is built on as the phases of the mission progress. The sharing of information could be as a result of direct communications between agents or through the use of the radio networks.

The individual scenarios show that the social networks are not particularly well distributed. However when the scenarios are amalgamated into one scenario, as may happen in a real life threat, it becomes a much denser network with better levels of participation. The network then becomes a split communications network with the Anti-Air Warfare Officer (AAWO) and Principal Warfare Officer (PWO) being the central nodes. Overall the analyses indicate that the Type 23 crew use a distributed network. Each crew member is connected (communication links) to other crew members and hence there are several channels with which communication can travel or information be shared.

Chapter 5 Case Study in RAF Boeing E3D Sentry

With contributions from Alison Kay, Mel Lowe, Paul S. Salmon, Rebecca Stewart, Kerry Tatlock and Linda Wells

This study was conducted on board a Boeing ED3 AWACS (Airborne Warning and Control System) aircraft in the Royal Air Force. The crew's role is as airborne surveillance, command and control, and weapons control and operations. They are essentially an aerial patrol that is able to provide information on enemy movement as well as providing control and the direction of friendly defensive and offensive airborne operations. The study used the EAST methodology to explore a complex communication system between eighteen team members and external agencies, where effective communication, decision making and coordination are essential to task success for this highly distributed communication network.

Five different missions were flown and observed. The missions were different, and due to the complexity of the task only individual crew members could be observed as opposed to the scenario being viewed as a whole. However an overall idea of communications could be ascertained from these individual observations.

Data are presented in the form of Social Network Analysis, Coordination Demand Analysis and Propositional Network Analysis. Shared situation awareness is also considered in this case study using the propositional networks. Knowledge objects are identified for the whole of the mission as well as for the phases of the mission. This gives an indication of where there is sharing of information. The networks also suggest where knowledge is built on as the phases of the mission progress. The sharing of information could be as a result of direct communications between agents or through the use of the radio networks or radar screens. Overall the analyses indicate that the E3 crew use a fairly distributed network. Each crew member is connected (communication links) to other crew members and hence there are several channels within which communication can travel or information can be shared.

Chapter 6 Case Study in Battle-Group HQ

With contributions from Dan Jenkins, Richard McMaster, Rebecca Stewart, Guy H. Walker and Linda Wells

These studies were conducted at the Command And Staff Training centre in Warminster. Military command and control relies heavily on tasks that require interaction with other team members, and where this is manifest, team working is principally concerned with the communication of information and development of Situation Awareness (SA). A relatively simple, yet robust, technological infrastructure underpins team tasks. It is heavily reliant on a combination of verbal communications and/or the translation of various planning 'products' into an integrated, collective, 4D spatial and temporal 'image' of the battle-space. It appears to be in this domain, based on the Communications Usage Diagram (CUD) method, that Network Enabled Capability (NEC) technology has much to offer. The assumption is that if the state of SA can be more rapidly and accurately acquired (and there seems little doubt that new technology offers this potential), then decision superiority can be achieved more quickly. If SA can also be shared in optimal ways throughout the system (which again, new technology appears to provide for), then unity of effort can also be achieved.

The Hierarchical Task Analysis (HTA) specifies how the configuration of people and technology changes in a task and context dependant manner. Three activity stereotypes are defined: semi-autonomous working, briefing and reviewing. The social network configures (and re-configures) itself numerous times during the enactment of military command and control (and the Combat Estimate specifically). As the network is re-configured, its constraints in terms of communications, density and centrality change. The design of NEC paradigms, therefore, is revealed to be more than just a consideration of technology in isolation. The specification of technology may be appropriate for one configuration, but inappropriate for another. The combination of HTA and SNA appears to provide one route into addressing this issue.

The knowledge base that underpins effective SA at the systems level changes in response to task phase, but also arises as a property of the constraining features of the configuration of people and technology. Systems level SA, at this summary level of analysis, appears to support, and be congruent with, task goals (as specified by the HTA).

In summary, the emergent properties associated with military (and indeed any) command and control scenario relate to the interplay between task, social and propositional networks.

Chapter 7   Development of a Generic Process Model of Command and Control

With contributions from Rob Houghton, Dan Jenkins, Richard McMaster, Paul S. Salmon, Rebecca Stewart, Guy H. Walker and Mark Young

Despite the differences in the domains, the command and control applications share many common features. First, they are typified by the presence of a central, remote, control room. Data from the field are sent to displays and/or paper records about the events as they unfold over time. Second, there is (currently) considerable reliance on the transmission of verbal messages between the field and the central control room. These messages are used to transmit report and command instructions. Third, a good deal of the planning activities occurs in the central control room, and these are then transmitted to the field. There are collaborative discussions between the central control room and agents in the field on changes to the plan in light of particular circumstances found in-situ. Finally, the activities tend to be a mixture of proactive command instructions and reactive control measures. It is hypothesised that one of the determinants of the success or failure of a command and control system will be the degree to which both the remote control centre and agents in-the field can achieve shared situational understanding about factors such as: reports of events in the field, command intent, plans, risks, resource capability, and instructions. This places a heavy reliance on the effectiveness of the communications and media between the various parties in the command and control system.

Analysis of the task analyses from these three domains led to the development of a taxonomy of command and control activities. The resultant data from the observational studies and task analyses were subject to content analysis, in order to pick out clusters of activities. These clusters were subjected to thematic analysis consistent with a 'grounded theory' approach to data-driven research. It was possible to allocate most of the tasks in the task analysis to one of these categories. To this extent, the building of a generic model of command and control was driven by the data from the observations and task analyses.

From development of a series of taxonomies (called receive, planning, rehearsal, communicate, request, monitor and review), and an analysis of the previous command and control models, it was possible to develop a generic process model. Construction of the model was driven by the data collected through observation from the different domains, and the subsequent thematic analysis and taxonomic development. In the tradition of 'grounded theory' the generic command and control model was as a result of our observations, rather than an attempt to impose any preconceived ideas of command and control. This may account for many of the differences between the current model developed in the course of the current research and those models that have come before it.

It is proposed that the command and control activities are triggered by events, such as the receipt of orders of information, which provide a mission and a description of the current situation of events in the field. The gap between the mission and the current situation lead the command system to determine the effects that will narrow that gap. This in turn requires the analysis of resources and constraints in the given situations. From these activities, plans are developed, evaluated and selected. The chosen plans are then rehearsed before being communicated to agents in the field. As the plan is enacted, feedback from the field is sought to check that events are unfolding as expected. Changes to the mission or the events in the field may require the plan to be updated or revised. When the mission has achieved the required effects, the current set of command and control activities may come to an end.

The model distinguishes between 'command' activities and 'control' activities. Command comprises proactive, mission-driven, planning and co-ordination activities. Control comprises reactive, event-driven, monitoring and communication activities. The former implies the transfer of mission intent whereas the latter implies reaction to specific situations.

Chapter 2

Modelling Command and Control

With contributions from Rob Houghton, Paul S. Salmon
and Guy H. Walker

Purpose and Scope

In order to situate the generic process model within its wider research context a review of the command and control literature has been conducted. Of particular concern to this review are the questions of what would define a useful 'model' of command and control. The literature provided several approaches to the challenge of modelling command and control, and the review addressed a number of broad questions, including:

- *Model metrics:* How are aspects of command and control measured or expressed either quantitatively or qualitatively?
- *Measures of modelling outcome:* How does the model define good or bad command and control system performance?
- *Degrees of model reconfigurability:* Is the model tied to a particular type of activity or situation? Is flexible enough to be reconfigured for use across a wide range of settings and contexts?
- *Construct validity and reliability:* Is the theoretical basis of the model sound?
- *Extent, nature and degree of dependency upon constraints and assumptions:* Are the assumptions the model is based on reasonable? Are the formal constraints within which the model falls appropriate or are they overly restrictive or too poorly specified?

These concerns cover the wider aim of extracting broad modelling trends from the prevailing literature, examples of paradigms and approaches, and the derivation of typologies and categories of outcome measure. The aim of the review, therefore, is not to provide an exhaustive account of every permutation of C4i related models but to illustrate the range and scope of relevant approaches and paradigms. Prior to presenting this review, the chapter will address generic issues associated with the problem of creating and using models.

Models

As Pew and Mavor (1998), in their report on the activity of the US National Research Council's *Panel on Modeling Human Behavior and Command Decision Making: Representations for Military Simulations*, the word 'model' can cover a host of definitions, from physical mock-ups to analytical descriptions. In Pew and Mavor

(1998) the term 'model' was taken to imply that '...human or organizational behavior can be represented by computational formulas, programs or simulations' (p.11). In this book, the notion of 'model' is much lighter, that is, our concern is to develop a useful description that can be applied, as a framework, to understand the operation of command and control systems. Thus, the intention is not to generate algorithms that can predict how a command and control system *should* operate, so much as to develop a framework in which to describe how such systems *do* operate. There are numerous modelling challenges underlying command and control. The difficulty contained by these challenges can perhaps be summarised by the fact that the 'real world is made from open, interacting systems, behaving chaotically' (Hitchins, 2000), and in the case of human actors, non-linearly. Complex systems like command and control also possess various real-time properties that cannot be considered 'designed' as such, they sometimes merely 'happen' (Hitchins, 2000). Therefore the notion of a commander representing something akin to the conductor of an orchestra is in some cases entirely false (Hitchins, 2000). Also, unlike neat linear systems the possibility exists (increasingly so with NEC) for there to be no clear boundaries between certain system elements, as well as no beginning and no end, given that goals are more or less externally adaptive. For the purposes of this book, a descriptive model serves four primary purposes: abstracting reality, simplifying complexity, considering constraint and synthesising results.

Abstracting reality

A model is an abstraction of reality (Wainwright and Mulligan, 2004), or 'a representation that mirrors, duplicates, imitates or in some way illustrates a pattern of relationships observed in data or in nature. ...' (Reber, 1995, p.465). A model is also a kind of theory, 'a characterisation of a process and, as such, its value and usefulness derive from the predictions one can make from it and its role in guiding and developing theory and research' (Reber, 1995, p.465). The purpose of a model is, therefore, to explain attendant facts, to characterise them, to represent the relationships between them, in a way that represents some form of direct analogue to the phenomena under analysis but in the most parsimonious way possible or that is appropriate.

Simplifying complexity

A model aims to simplify complexity. Complexity is related to the amount of information needed to describe the phenomena under analysis. The closer that the phenomena under analysis approaches complete randomness, the more data are needed until it 'cannot be described in shorter terms than by representing the [phenomenon] itself' (Bar Yam, 1997). However, 'Something is complex if it contains a great deal of information that has a high utility, while something that contains a lot of useless or meaningless information is simply complicated' (Bar Yam, 1997 cited in Grand, 2000, p.140). The primary purpose of the current generic model of C4i is to reduce complexity and particularly when partnered with additional modelling techniques, to offer outcome metrics that can detect and describe non-random emergent properties. Emergent properties exist where the 'characteristics of the whole are developed (emerge) from the interactions of their components in a non-apparent way' (Bar Yam, 1997). Previous analyses of live C4i scenarios using the EAST methodology have detected the presence of emergent properties related to task, social and knowledge networks and in so doing demonstrate that information contained within these scenarios may indeed be complex, but is far from random. The optimal model of command and control can be defined as one that is sensitive enough to detect cogent emergent properties, whilst containing merely 'sufficient' complexity to explain (and predict) these 'widely observed properties and behaviours in terms of more fundamental, or deeper, concepts' (Wainwright and Mulligan, 2004; Builder, Bankes and Nordin, 1999). These apparently simple requirements are heavily tempered by the fact that:

> the world being modelled has an inviolable nature; it cannot be exhaustively described. We can model the world but we can always go back to find new perspectives for describing what we are modelling, usually involving new perspectives on what constitutes information (data), new languages for modelling, and new perspectives on the purpose for constructing models. (Clancey, 1993, p.41)

Considering constraint

A crucial aspect of any modelling enterprise is the quality of the constraints and assumptions within the model. For a model to have scientific viability it must feature some variety of formal constraint. In simple terms, constraints are the limits on what a model can and cannot do and should (for the most part) correspond with the real world. For example, if we were modelling some aspects of human manual labour, this would naturally dictate a constraint that within the model any one individual has at maximum two arms and two legs. A model may be otherwise impressive in any other aspect but if its constraints are incorrect or inappropriate then the rest of the model is invalid. A model may even appear to give the 'right' answer if treated as a 'black box' model where we concern ourselves only with the input and the output, but if it requires the representation of impossibilities (for example, three armed radio operators, speed of light reaction times and so on) to generate those results it is questionable whether the model is of genuine theoretical value: a good model 'plays by the rules' so to speak. Whilst it might be obvious how many limbs to ascribe to an individual, less concrete variables are more difficult to accurately and uncontroversially constrain; things like attentional capacity, situation awareness and teamwork. Whilst these are essential constructs in the conceptualisation of command and control, they have long suffered from problems in their precise characterisation (and in some cases possibly always will given that mental qualities like attention and awareness are possibly products of reification). Furthermore, where models are themselves particularly complex or abstract by nature of their operating principles (for example, a statistical Hidden Markov-Chain Model) there can be additional challenges in translating realistic constraints into parameters (for the author) but also assessing the constraints (for the reader); not all models are so transparent that one might easily relate, say, a set of arrows on a graph or a set of numbers in a vast matrix with an inappropriate 'take' on the real world.

Synthesising results

Chapters 4, 5 and 6 contain a series of studies conducted in a wide range of domains. A generic model allows the results from these studies to be combined into a coherent account, and the model itself provides a motivation for the analysis, that is, if the results do not bear out the relationships implied by the model, then it needs to be modified. The model would also provide a means for comparing and contrasting different approaches to command and control across the different domains, and also allows analysts to tease out those factors that seem to be contextual and those that are more generic aspects of performing command and control activity.

Specification

Defining command and control

In this book, we follow NATO (1988) in separating the concept of 'command' from that of 'control'. For NATO (1988), 'command' is '…the authority vested in an individual…for the direction, coordination and control of military forces'. This implies that an individual will be given the role of Commander, that this individual will (through this role) be imbued with sufficient authority to exercise command, and (by implication) this command will involve defining the goal (intent, effect) that Forces under the individual's command will achieve.

Builder, Bankes and Nordin (1999) follow NATO (1988) in their definition of command and control:

> Command and control: The exercise of authority and direction by a properly designated [individual] over assigned [resources] in the accomplishment of a [common goal]. Command and control functions are performed through an arrangement of personnel, equipment, communications, facilities, and procedures which are employed by a [designated individual] in planning, directing, coordinating, and controlling [resources] in the accomplishment of the [common goal]. (p.11)

Combining command (authority) with control (the means to assert this authority) leads to 'unity of effort in the accomplishment of a [common goal]' (Jones, 1993, p.2). Despite the militaristic undertones, the notion of command and control is itself generic. The separation of activity associated with command from that of control has a number of useful benefits. For example, in military parlance, 'Mission Command' involves passing a Commander's Intent down a chain of command in such a way as to allow lower levels to elaborate and develop the command in the light of contextual demands. Thus, a Commander might have an Intent to 'secure a route from town A to town B'. This might be interpreted by Officers in the field as (a) patrol the road between the towns; (b) secure a bridge along the route; (c) repel any attacks on the bridge; and (d) disrupt any attempts to damage the road. Each of these activities could be passed to specific Units, under the command of an Officer who will then define these activities according to contextual demands, for example, activity (a) could be performed according to different time-intervals, using different resources, with greater or lesser shows of strength etc. In a similar manner, the notion of Sectorisation in the Fire Service requires Officers to divide activities in major incidents between different teams, for example, managing water supply, using breathing apparatus. Each 'Sector' exercises different skills under the leadership of different Officers. Taken together, these examples point to the exercise of Command as the broad definition of a goal and the provision of appropriate resources to achieve that goal, whereas Control is more concerned with the management of these resources in the ongoing pursuit of that goal in the light of changing contextual demands.

Generic properties of command and control

Given the definitions above the (relatively) invariant properties of command and control scenarios can be distilled down to the following three features:

- A common overall goal (this may, however, be comprised of different but interacting sub-goals).
  – Corollary: systems of command and control are goal-oriented systems.
- Individuals and teams acting individually or in unison.
  – Corollary: there is the need to coordinate activity.
- Teams and sometimes individuals are often dispersed geographically.
  – Corollary: there is the need to communicate and share 'views' on the situation

Beyond the descriptive level, command and control by definition is a collection of functional parts that together form a functioning whole. Command and control is a mixture of people and technology, typically dispersed geographically.

In their consideration of command and control, Alberts and Hayes (2006) consider the SAS-050 'cube' space of possible command and control structures. This assumes three broad Independent Variables,

- Distribution of information – this could be sent from one person to another, or could be broadcast to all members of a network;
- Patterns of interaction – this could take the form of a top-down, hierarchical command structure or could take a more open, or 'distributed' form of management;
- Allocation of decision rights – this could have Intent originating from a single source, for example, a Commander, or arising from some form of 'democratic' decision making.

Figure 2.1 shows how these three Independent Variables can be mapped onto a cube. It is possible to place different forms of command and control structure in this cube. Thus, for example, a 'traditional' hierarchical command structure could be located in the bottom left-hand corner, whereas a 'power to the edge' structure could be located in the top right-hand corner. Other forms of command and control structures could be located throughout the cube.

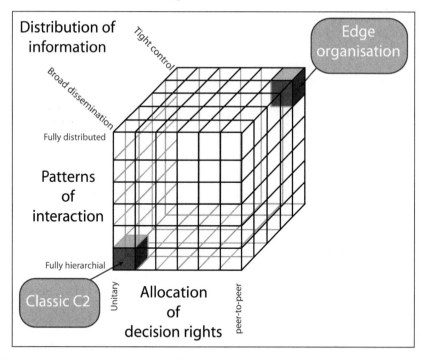

Figure 2.1 SAS-050 Space of C^2

Network Enabled Capability

Network Enabled Capability (NEC) is a way of 'doing' command and control. It is a term used to describe what is at present a nascent paradigm within British military command and control, but also an extremely cogent modelling issue. It has been described thus: 'NEC is about the coherent integration of sensors, decision makers, weapons systems and support capabilities to achieve the desired effect' (MoD, 2005a). Diverse command, reconnaissance, weapons, support and decision making assets will be interlinked by way of a 'network of networks', which will deliver and transfer data and communications in a flexible manner. In essence this is an information-based approach to command and control; aside from their obvious operational capabilities, assets also take the role of active producers and consumers of information. Expected benefits include:

- Increased provision of timely information allowing rapid response (ideally to the point where the opposing force's own decision loop is undercut).
- Improved interoperability (across domains, services, agencies, nations etc.).
- Increased tempo of operations.
- The development of more effective command and management structures (likely to be marked by a reduction in or flattening of existing hierarchies and increasing freedoms given to commanders lower in the chain of command).

- Shared situation awareness across actors (that is, 'singing from the same hymn sheet') that will in turn allow the emergence of the so-called synchronisation effect. In other words self-coordinating group actions will be enabled by improved information about what colleagues intend to do, are doing and have done, and the operational picture they are responding to.

Success in achieving these aims appears to be contingent on a set of central presuppositions about the nature of technology and human cognition: that shared information can be *actively* shared through networking; that shared information could become in the mind of operators shared knowledge (that is, it is meaningful and supports action); shared knowledge in turn leads to shared situation awareness; that shared situation awareness enables synchronisation and that ultimately synchronisation results in operational effect.

In essence NEC represents the application to command and control of principles and technologies of the so-called information age. In particular, the effect of the Internet and related technologies within the commercial sphere to increase the responsiveness and flexibility of businesses appears to have been a primary inspiration (Kaufman, 2004).

Structural Models of Command and Control

The dominant perspective – the cybernetic paradigm

C4i is variously defined as a form of 'management infrastructure [...] for any [...] large or complex dynamic resource system' (Harris and White, 1987). It is immediately apparent based on this review that the dominant modelling perspective, the so-called 'cybernetic paradigm' (Builder et al., 1999) accords the structural aspect of command and control particular prominence. Dockery and Woodcock (1993) are unequivocal in stating that, 'Since [command and control] [...] concerns itself with providing structure to combat, its description should be in terms of structure' (p.64).

Under this structural/cybernetic paradigm a command and control scenario is divided into linked functional parts that exchange and modify signals that can be specified according to mathematical formulae (Builder et al., 1999). Or as similarly expressed from a systems dynamics point of view, 'the mathematical modelling of an assemblage of components so as to arrive at a set of equations which represent the dynamic behaviour of the system and which can be solved to determine the response to various sorts of stimuli' (Doebelin, 1972, p.4). An example of these 'equations', and their relation to the dynamic behaviour and structure of command and control 'systems' are reproduced in Table 2.1 from Dockery and Woodcock (1993) merely as an illustration of the outcome measures that can be derived.

Table 2.1 Sample of mathematical metrics for key constructs in command and control (from Dockery and Woodcock, 1993, p.66)

Mathematical Tool	Modelling Requirement
Catastrophe Theory	To capture in both a time independent and time dependant manner the global non-linear responses of combat in terms of a few control variables, therefore essentially presenting the commander's command and control perspective on the battle.
Category Theory	To define measures of effectiveness and to embed the whole command and control modelling process in a larger context.
Cellular Automata	To capture the self-structuring of combat based on minimal nearest neighbour rule sets, essentially to present the small unit command and control perspective of combat.
Chaos Theory	To capture the observed chaotic dynamical nature of combat and to identify attractors and to present the combat evolution perspective.
Fuzzy/Partial/Stochastic Differential Equations	To describe attrition based combat processes.
Entropy Computation	To capture the effects of casualties upon the structure of the fighting.
Fractals	To capture deployment hierarchies.
Fuzzy Sets	To capture the imprecision in all phases of the military command process from data to orders.
Relativistic Information Theory	To capture the idea that organisations move relative to each other in a metaphorical sense of internal organisational efficiency.
Langrangian Formulation	To define classical, albeit heuristic, expressions such as combat momentum (tempo).
Petri Nets	To express the transactions that occur in a command and control system between its elements.
System Dynamics	To capture feedback and feed-forward aspects of command and control interacting with combat processes.
Games Theory	'Ways in which strategic interactions among rational players produce outcomes with respect to the preferences […] of those players, none of which might have been intended by any of them' (Ross, 2005).

Lawson's model of command and control

Under Lawson's model (1981), command and control can be viewed as an information processing chain with data flowing between the environment, one's own forces and the command centre. The model in Figure 2.2 epitomises this perspective. The model is rooted in the idea that there is some desired state that the command centre seeks to achieve. Data are extracted from the environment and processed. The understanding of these data are then compared with the desired state. If there is any discrepancy between the desired state and the current state, the command centre has to make decisions about how to bring about the desired state. These decisions are turned into a set of actions, which are then communicated to their own forces. The data extraction cycle then begins afresh.

Lawson's model owes much to the ideas of control theory. The comparison of actual and desired states implies a feedback process and some form of regulation. Central to his model, therefore, would be the 'compare' function. The feedback involves control of 'own forces' to affect a change to the environment. The notional 'actual' and 'desired' states imply phenomena that can be described in terms of quantitative, discrete data; in other words it is not easy to see how the model would cope if the actual state was highly uncertain. Nor is it easy to see what would happen if the changes to the environment led to consequences which lay outside the limits defined by the discrete state. The model does indicate the central issue that command can be thought of as working towards some specified effect or intent but suffers, however, from its apparent reliance on a deterministic sequence of activities in response to discrete events.

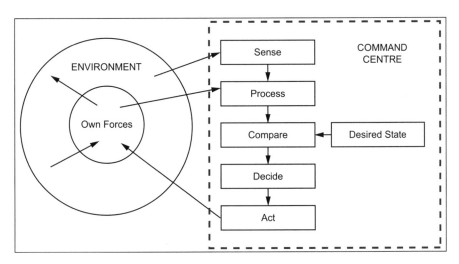

Figure 2.2 An adapted version of Lawson's model of the command and control process

The OODA Loop

One of the most commonly used structural models of Command and Control is referred to as the OODA Loop. This provides a tactical-level perspective on Command and Control as a process involving Observation (of events in the field), Orientation (of resources to deal with those events), Decision (of how best to manage those resources in order to deal with those events, under some set of constraints), and Action (the use of resources to deal with the events). Figure 2.3 shows an example of the OODA Loop, together with some of the factors that could have a bearing on performance.

The Headquarters Effectiveness Assessment Tool

Another example derived from the 'cybernetic approach' is the Headquarters Effectiveness Assessment Tool (HEAT). HEAT's *raison d'etre* is to provide an objective measure of headquarters effectiveness based on the premise that effective headquarters approach command and control activities in quantitatively different ways than ineffective headquarters (Choisser and Shaw, 1993). The HEAT method, like Lawson's model above, is based on (mathematical) normative systems and optimal control theories, whereby the 'objective of the [commander] is to determine control activities that will induce the evolution of the system towards an acceptable goal' (Choisser and Shaw, 1993; p.48). The command and control system, in common with Lawson's model, is viewed as a variation on a deterministic closed loop system, expressed in very simple terms in Figure 2.4. The idea expressed in the HEAT methodology/theory is that rational command and control systems will possess some of the anticipatory and self-optimising properties of a normative system. In a practical sense HEAT is used within war-game simulations and various mathematically derived outcome measures are derived from the underlying control (and decision making) model.

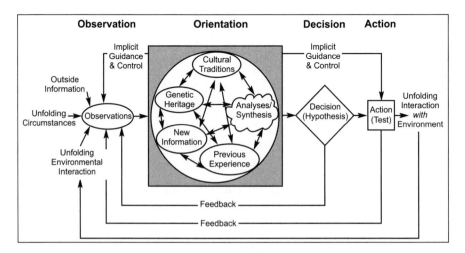

Figure 2.3   The OODA loop

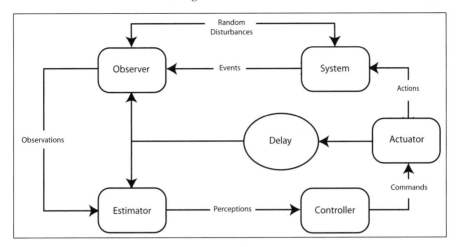

Figure 2.4   Example of closed loop feedback control system (from Choisser and Shaw, 1993)

The Network Centric Operations Conceptual Framework

The Network Centric Operations Conceptual Framework (NCO CF) comprises four main elements. Information sources, which could be sensors, people, or any other originator of intelligence; value-added services, which could involve the integration of information from several sources; the Command and Control function; and Effectors, which are the means by which actions are performed. Figure 2.5 shows a schematic of the NCO CF process model. Notice how the intention is to contrast the activity relating to Individuals (on the left-hand side) with that of Groups (on the right-hand side), with a central role for 'Quality of Interactions'. This figure can be read as an elaboration of the previous structural models, in that its focus is on higher-level 'quality' issues and their interactions.

Issues in structural modelling

The models that arise from this dominant perspective all tend to approximate in varying degrees towards control theory models and as Builder et al. (1999) put it, could just as easily apply to a thermostat as it could to a C4i system. Although practitioners of the cybernetic paradigm would argue that much of the complexity lies 'behind' such simple systems (within the mathematical metrics expressed in Table 2.1, for example), the fact, at bottom, still remains. Overall, the specific outcome metrics from these models are relatively hard to divine, but examples include 'catastrophe manifolds', 'butterfly landscapes' (a form of five dimensional graph), 'control coefficients', 'time dependant behaviour', etc. These mathematical metrics provide measures related to overarching 'mission based' constructs such as survivability and effectiveness. Changes in these overarching constructs are based

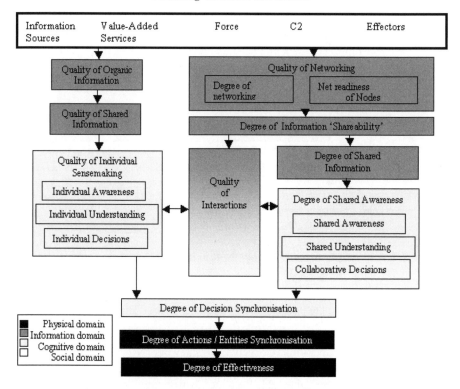

Figure 2.5 Network Centric Operations Conceptual Framework (adapted from Alberts and Hayes, 2006)

on changes to (and the mathematical interplay of) independent variables such as force strength, fire power, decision aids, etc. (Dockery and Woodcock, 1993).

The attractiveness of, yet inherent danger associated with, this modelling perspective is perhaps hinted at by Doebelin (1972), who states, 'From our intuitive ideas about the cause-and-effect nature of the physical world, it is clear that if we precisely define a model of a physical system and subject it to specific known inputs, the outputs are completely determined' (p.4). The problem is that the robust relationship between cause and effect does not necessarily hold true for systems involving human cognition as it might do for systems involving merely physical entities with known properties and known input-output characteristics (Rasmussen, Pejtersen and Goodstein, 1994). Simply put, cybernetic models 'inadequately represent the complex and idiosyncratic activities of humans in [command and control]' (Builder et al., 1999). 'When the [structure] is put to work, the human elements change their characteristics; they adapt to the functional characteristics of the working system, and they modify system characteristics to serve their particular needs and preferences' (Rasmussen et al., 1994). 'It is the un-specifiable messiness of the neural system – becoming organised in new ways at the time of the interaction itself – which gives human behaviour its robust, always adaptive character' (Clancey, 1993, p.41). In modelling terms two strategies within the cybernetic paradigm can be implied as routes taken to attempt to overcome this inconvenience. Firstly, the human is simply subsumed into a complete physical system, where any non-linearitys are catered for using (ever more) complex mathematical techniques (for example, Dockery and Woodcock, 1993). Secondly, the complex role of the human may be recognised, but is still ultimately reduced and overlain on to an underlying control theoretic model (Choisser and Shaw, 1993; Edmonds and Moss, 2005; Levine, 2005).

Summary of the cybernetic paradigm

In summary, the issue of model constraints appears to be felt particularly acutely and while cybernetic models might provide a robust basis for understanding control, they appear to be restricted in their ability to model command. This is simply because the models tend to rely on the assumption that the role of C2 is to react to changing events in the world. Thus, the models are 'event-driven' and reactive rather than proactive or anticipatory.

Network Models

Introduction

If taken as a form of doctrine the structural/cybernetic perspective yields several serious limitations for modelling the multi-faceted nature of command and control. If, however, the notion of structure is assumed to be a component of a system (as opposed to a complete characterisation), then structural aspects of command and control can be modelled in several alternate and useful ways.

From the perspective of Organisational Theory an organisation can be defined as 'a collection of interacting and interdependent individuals who work toward common goals [...]' (Duncan, 1981). Here the focus is directed onto 'individuals', and their links and interrelations. The straightforward organisational chart is a simple example of this and 'shows the relationship between specific jobs or roles within [an] organisation' (Arnold, Cooper and Robertson, 1995, p.2). Put more explicitly, an organisational chart represents the links between roles, where the hierarchical organisation of roles is reflective of command (and the cybernetic paradigm), and the links that exist between roles reflective of control (and the communications between individuals).

Example: Hitchen's N-squared chart

This notion of roles and links is related to a simplistic descriptive model of military command and control presented by Hitchens (2000) in Figure 2.6.

In Hitchens' N-squared (N^2) chart, information intersects pairs of roles (shown in capitals on the diagonal). The example given is militaristic, where the content of information flow downwards from the commander is represented in the upper quadrant and information flowing upwards to the commander in the lower quadrant. The commander, therefore, interchanges 'decision' information to the operations

COMMANDER	Tasking	Decisions	
Enemy Intentions	**INTELLIGENCE**	Enemy Intentions	Needs
Operations Plans	Operations Plans	**OPERATIONS**	Needs/Priorities
	Constraints	Constraints	**LOGISTICS**

Figure 2.6   The N-squared (N^2) chart from Hitchens (2000): A simplistic generic model of command and control

role. Similarly, the logistics role interchanges information on 'constraints' to the operations role. And so on. In basic terms this simplistic model expresses the links between roles and the broad topic of information that the links are facilitating.

Example: Hierarchical Task Analysis

Hierarchical Task Analysis (HTA) can also be viewed, perhaps unconventionally as a form of network model. Task analysis is the activity of collecting, analysing and interpreting data on system performance (Annett and Stanton, 2000; Diaper and Stanton, 2004). It is one of the central underpinning analysis methods within the DTC HFI's EAST method alluded to earlier. According to Stanton (2004), task techniques can be broadly divided into five basic types: hierarchical lists (for example, HTA and GOMS), narrative descriptions (for example, the Crit and Cognitive Archaeology), flow diagrams (for example, TAFEI and Trigger Analysis), hierarchical diagrams (for example, HTA and CCT), and tables (for example, Task-Centred Walkthough, Interacting Cognitive Subsystems-Cognitive Task Analysis: ICS-CTA, HTA, Sub-goal template: SGT, and Task Analysis For Error Identification: TAFEI). Some methods have multiple representations, such as HTA, which can be viewed as a hierarchical text list, a hierarchical diagram or in tabular format.

HTA is often referred to as a means to 'model' an interaction. The emergent properties are related to functional groupings of tasks within a hierarchy (which is not readily apparent in most cases), and the reductions in complexity that arise from structuring tasks in this way. HTA permits multiple modelling perspectives; at the bottom layer the entire task is represented in full, whereas higher levels within the hierarchy represent progressively more parsimonious descriptions. Either may be more or less appropriate to the type of analysis required. HTA is a more rigorous means of describing a system in terms of goals and sub-goals (as opposed to roles and sub-ordinates) and departs from an organisational chart in terms of its more rigorous and highly defined internal logic used to create a nested hierarchy, and also in the provision of rules and feedback that define the deterministic (cause and effect) enactment of tasks contingent upon specific external conditions being met.

Example: Social and propositional networks

Various other deterministic approaches, such as pyramid laws and tree structure models can be used as ways to offer predictive network based modelling. Examples of metrics include two defined by Hitchens (2000, p.6) based on pyramid laws, namely; diminishing lateral communication ('it is harder to get agreement with someone who is more distant laterally in the hierarchy'), and vertical data compression ('as you go up the [hierarchy] data gets compressed at something like the span of control [width of the pyramid] at each level'), as shown in Figure 2.7.

Similar numerical metrics can be derived using social network analysis (Driskell and Mullen, 2005). Social networks and organisations (according to their various definitions) are virtually isomorphic, as are various numerical metrics derived from deterministic network approaches. For example, the metrics derived from social network analysis map onto those derived from pyramid laws; diminishing lateral communication is related to the property of 'degree' (the number of other positions in the network in direct contact with a given position), whereas vertical data compression is related to the property of betweeness (the frequency with which a position falls between pairs of other positions in the network). In a similar way to pyramid laws, early research on social networks illustrated how basic structural properties of a network can influence the outcome measure of efficacy of communications within the structure

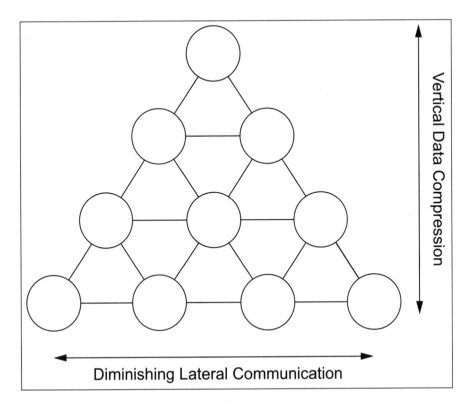

Figure 2.7   Pyramid structure metrics

(Figure 2.8). For example, the 'cross' prototype is advantageous for information integration but can also lead to information overload for the central node.

A further relevant expression of a network model is a Propositional Network. Propositional Networks are similar to semantic networks in that they contain nodes (with words) and links between nodes. It is argued that the application of basic propositions and operators enables dictionary-like definitions of concepts to be derived (Ogden, 1987). Stanton et al. (2006) take this basic notion and extend it to offer a novel way of modelling knowledge in any scenario. Knowledge relates strongly to the concept of situational awareness. Situational awareness is about 'knowing what is going on' (Endsley, 1995) and enables decisions to be made in real time, and for the 'commander' in a command and control scenario to be 'tightly coupled to the dynamics of [their] environment' (Moray, 2004, p.4). *This* is a large part of how decision superiority is achieved. A systems view of SA (and indeed an individual view as well) can be understood as activated knowledge (Bell and Lyon, 2000) and, therefore, propositional networks offer a novel and effective means of modelling this 'systems level' view of SA.

A major advantage of propositional networks is that they do not differentiate between different types of node (for example, knowledge related to objects, people or ideas) so that from a modelling perspective they are not constrained by

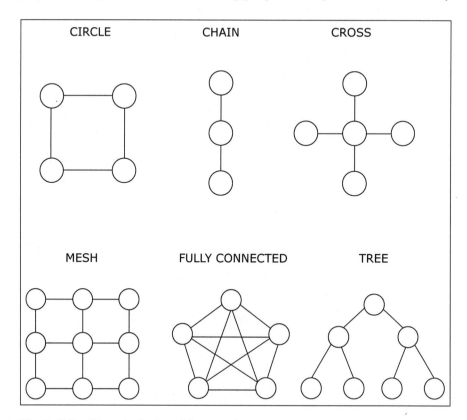

Figure 2.8   Prototypical social networks

existing structures of people and objects, rather to the required knowledge elements associated with a scenario. It is possible to model the temporal aspects of SA by animating the propositional network in terms of active and non-active knowledge objects. To do this the scenario is divided into task phases allowing active and non-active knowledge objects to be specified and represented.

In summary, propositional networks enable SA (at the systems as well as individual level) to be modelled in command and control scenarios and for the emergent property of knowledge activation, structure and (systemic) SA to be measured. Figure 2.9 illustrates a propositional network and Table 2.2 illustrates the metrics applied to the network to reveal the emergent property of SA as it relates to 'key aspects of knowledge'.

Table 2.2 Propositional network metrics to detect emergent property of SA in relation to key knowledge objects

	Scenario				
Knowledge Object (KO)	**Shift Hand-Over**	**Departure**	**Over-Flight**	**Holding**	**Approach**
Pressures	■	■			■
Runway	■	■			
Radar	■	■	■	■	■
Stack	■	■		■	■
Aircraft	■	■	■	■	■
Flight Level	■	■	■	■	■
Strip	■	■			
Position	■		■	■	■
Read-back		■	■		■
Instruction		■	■	■	■
Acknowledge		■	■	■	■
Call Sign		■	■	■	■
Flight Plan		■	■	■	■
Plan		■	■		■
Options		■	■		
Sector		■			
TOTAL KO (16)	**8**	**10**	**10**	**7**	**8**

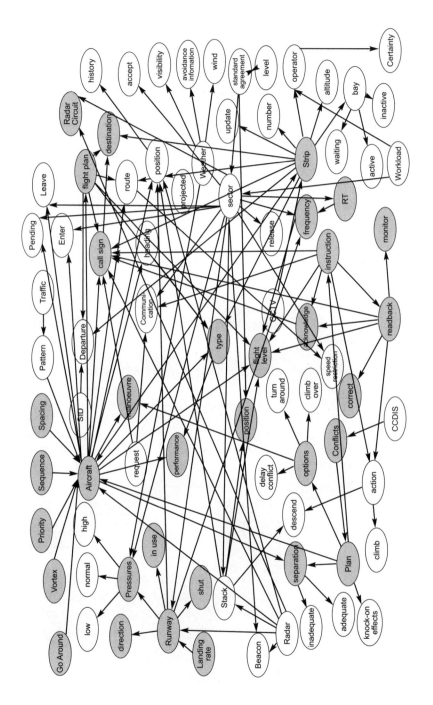

Figure 2.9 Propositional network from the air traffic control work domain

Summary of network approaches

Network models cannot be said to be widely applied in command and control settings. The various approaches do, however, appear to be highly appropriate for modelling a range of emergent properties such as task structure, social organisation and situational awareness. This approach is currently the basis for the DTC HFI's EAST methodology. While showing promise for modelling various emergent properties of specific C4i scenarios the challenge still exists in terms of bringing together these disparate networks and properties into a predictive model. This appears to be feasible. By linking various networks it is possible to subject individual networks to specified changes and to model the propagation of these changes across other linked networks. For example, modifying the social network may predict changes in terms of task structure and knowledge. This falls within the purview of current and future work being undertaken within the DTC HFI.

Dynamic Models

One approach to using dynamic models to assess the structure of military organisations is the FINC (Force, Intelligence, Networking and C2) methodology described by Anthony Dekker (Dekker, 2002). This approach considers the actions of organisations in terms of the deployment of Force, the gathering, fusion and communication of Intelligence, the extent of Networking and the number and role of C2 units.

Dekker (2002) tested the performance of different command structures in playing a simplified and abstracted war-game called SCUD Hunt in which players allocate force and intelligence assets within a 4 x 4 board in order to ultimately hunt down and destroy hidden SCUD missile launchers. On the basis of intelligence, air strikes can be called in on squares on the board. However, air strikes are not instantaneous with target detection by intelligence assets; intelligence and, in turn, orders to initiate action must be passed up and down the command structure. Thus a command structure that places many intermediary units between force, C2 and intelligence is one that is likely to be quite sluggish in response as there is a time delay encountered each time a message must be relayed. On the other hand, command structures with more intermediary units usually build in a high level of connectivity. In turn this means that intelligence be pooled and thus the accuracy of that intelligence is ultimately increased.

Within this paradigm, experimental manipulations were also made by Dekker, one to vary the reliability of sensor data (thus varying the importance of fusing intelligence) and the other to vary the speed at which targets changed locations (this therefore acted as an indirect measure of tempo for the SCUD Hunters; for example, a slow tempo command structure would get few if any hits against fast moving targets as it would not be able to respond quickly enough). Performance in the game is measured through the number of SCUDs destroyed, the number of SCUDs missed, and the number of false alarms.

Whilst SCUD Hunt was originally designed as a game to be played by humans in a laboratory setting, Dekker wrote a piece of software which carried out thousands of automated statistical trials in which different command network configurations repeatedly and automatically 'fought out' a game of SCUD Hunt. This so-called Monte Carlo approach to simulation allows the quantitative assessment of systems that have been represented probabilistically but are too complex (that is, have too many interacting degrees of freedom) for analysts to directly assess them otherwise.

Command structures used in SCUD Hunt

The following eight basic command structures were evaluated by Dekker (2002) within the SCUD hunt paradigm.

Centralised architecture without information sharing
Within this simple network architecture we see that intelligence data from four intelligence assets is collated by a central Intelligence HQ unit and passed on to a Strike HQ unit, which finally directs the attacks of four strike assets. This command structure is associated with the USAF (United States Air Force) who have good communications, good intelligence (from AWACS aircraft) and can, owing to the inherent speed of jet aircraft, move force assets into position rapidly. It is a fairly hierarchical network in which subordinates answer to superordinates and there are no direct links between strike and intelligence assets; information flows via the chain of command itself.

Split architecture without information sharing
This is very similar to the foregoing centralised architecture, the only modification being the addition of intermediary layer C2 units (Wing A and Wing B) between Strike HQ and the Force asset squadrons themselves. This architecture is more common in land based operations where benefit is derived from having local command units owing to issues like the complexity of terrain. As compared with the Centralised architecture there is a clear cost paid for this extra command layer in that it adds an extra delay between orders being issued by the HQ getting to the Force squadrons.

Distributed architecture without information sharing
The distributed architecture contrasts strongly with the centralised and split forms; as can be seen in Figure 2.12, each Intelligence and Force asset is tied together via a single distributed HQ C2 unit. Thus there are in essence four autonomous self-contained armies with their own intelligence and strike assets in the field. This architecture is most often found in the context of special operations where decision making must be done rapidly with regard to small-scale actions. Alternatively, it also describes a 'cell structure' used by terrorists and covert intelligence operatives. The self-contained nature of the groupings means that the destruction or infiltration of the unit has its impact restricted to that unit. Clearly one disadvantage of this approach is that information is not shared outside each autonomous grouping.

Modelling Command and Control 27

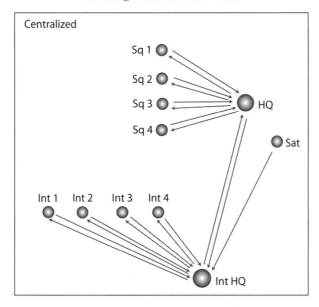

Figure 2.10 Centralised architecture without information sharing. The topology of the network shown in Figure 2.10 is similar to the 'cross' topology in Figure 2.8, with the HQ/Int HQ having central positions and nodes branching off

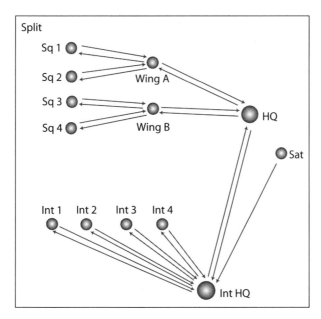

Figure 2.11 Split architecture without information sharing. The topology of the network shown in Figure 2.11 is similar to the 'tree' topology in Figure 2.8, with nodes branching off HQ

Negotiated architecture without information sharing

The negotiation architecture is quite similar to the distributed architecture, the only change being that now C2 HQ units can communicate with each other to share information. This 'peer to peer' style arrangement is commonly found with regard to emergency services (according to Dekker), as each unit will tend to cover a geographical area and work within that area whilst communicating with peers in other areas.

Architectures 5 to 8 inclusive: '...with information sharing'

Within his original report Dekker also added four 'information sharing' versions of the four command structures already described wherein intelligence is disseminated from intelligence assets to all other C2 HQ units. In the case of centralised and split architectures this does not change the physical layouts of the architectures, just alters their operations by adding an extra degree of delay to processing in the intelligence HQ (in the '...without information sharing' variants it is assumed the intelligence HQ relayed in parallel four packets of intelligence data to the strike HQ; with information sharing there is an extra time delay whilst the intelligence inputs are fused together). In this case of the distributed and negotiated architectures this means additional connections between intelligence and HQ units (see Figures 2.12 and 2.13). These two variants represent the new paradigm of Network Enabled Capability in which intelligence is shared within a densely interconnected network of sensors and communication links.

Analysis of performance

The basic measures that Dekker used in his Scud Hunt approach are delay analysis and intelligence analysis. Variations on the time taken for messages to pass from one node in the network to another, that is, measures of delay, leads to an Information Flow Coefficient, which shows how well a given structure can pass information from sensor to shooter assets in order to complete a task. A similar delay can be measured in the exchange of information between shooter assets, for example, to prevent multiple attacks on a target, and this is the Coordination Coefficient. If one applies a simple decay function to information, then the longer it takes to move through the network, the less 'up-to-date' it will be and (hence) the lower the Intelligence Coefficient.

Table 2.3   Relating SAS-050 variables to Scud Hunt Coefficients

SAS-050 Variable	Scud Hunt Coefficient	Explanation
Distribution of Information	Intelligence	Affected by Timeliness
Allocation of Decision Rights	Coordination	Affected by Synchronisation
Patterns of Interaction	Information Flow	Affected by C2 Structure

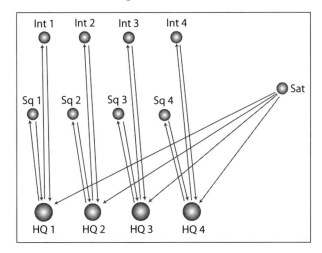

Figure 2.12 Distributed architecture without information sharing. The topology of the network shown in Figure 2.12 is similar to the 'cross' topology in Figure 2.8, with the Sat having central positions and nodes branching off

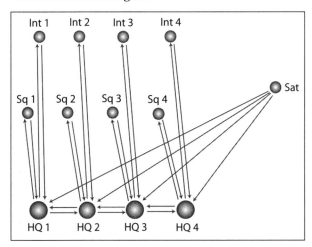

Figure 2.13 Negotiated architecture without information sharing. The topology of the network shown in Figure 2.13 is similar to the 'mesh' topology in Figure 2.8

From Table 2.4, it is possible to make a mapping between the models implied by the SAS-050 cube (see Figure 2.1) and those used by Dekker in his Scud Hunt studies. The mapping is not as complete as one might hope, there is not always a one-to-one correspondence between models, and that is likely to be due to the differences in emphasis placed on the concept of command and control between the two reports. For example, Dekker was primarily concerned with delays in information-flow in the task of linking a sensor to a shooter, whereas the SAS-050 report was concerned with more general aspects of command and control (including sense-making, shared awareness and coordination). However, there is enough common-ground between the assumptions underlying Dekker's models and the definitions provided by SAS-050 for this exercise to be useful. The mapping is shown in Table 2.4. The numbers in the SAS-050 model column refer to each corner of the cube (starting with 1 in the bottom left and moving clockwise around the cube).

Table 2.4   Mapping Dekker's SAS-050 models

Coord.	Intel.	Info.	Dekker Model Network	Pattern of Interaction	Allocation of Decision Rights	Distribution of Information	Process	SAS-050 model
4	8	1	2. Split	Fully hierarchical	Peer-to-peer	Tight control	Broadcast sensor data; orders to specific node	2
7	6	2	6. Split, share	Fully hierarchical	Peer-to-peer	Broad dissemination	Broadcast sensor data; broadcast orders	6
8	4	3	5. Cent, share	Fully hierarchical	Unitary	Broad dissemination	'Sensor' to 'command'; broadcast orders	5
6	7	4	1. Cent	Fully hierarchical	Unitary	Tight control	'Sensor' to 'command'; orders to specific node	1
5	2	5	8. Neg, Share	Fully distributed	Peer-to-peer	Broad dissemination	Broadcast sensor data; broadcast orders	8
3	1	6	7. Dist, Share	Fully distributed	Unitary	Broad dissemination	'Sensor' to 'command'; broadcast orders	7
2	5	7	4. Neg	Fully distributed	Peer-to-peer	Tight control	Broadcast sensor data; orders to specific node	4
1	3	8	3. Dist.	Fully distributed	Unitary	Tight control	'Sensor' to 'command'; orders to specific node	3

Modelling Command and Control 31

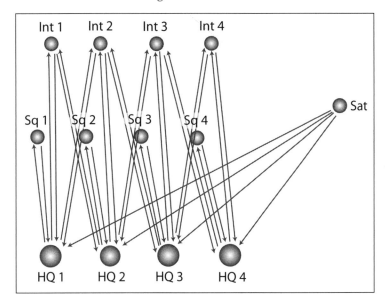

Figure 2.14 Distributed with information sharing. The topology of the network shown in Figure 2.14 is similar to the 'fully-connected' topology in Figure 2.8

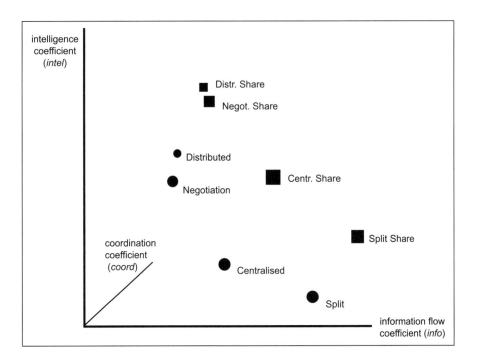

Figure 2.15 Performance of different network structures

The columns to the left of Table 2.4 show the approximate ranking of Dekker's models (from Figure 2.15) on the coefficients of Coordination, Intelligence and Information. The obvious implication is that (according the metrics he applied), the 'best' networks for one coefficient are not necessarily the best for another. This indicates that the 'optimal' network will depend on the type of activity being performed.

For Intelligence, optimal performance tends to arise when 'sharing' occurs (irrespective of network structure). This would appear to offer some support for what might be considered the more extreme views of NEC, for example, 'power to the edge' (which is possibly '8. Neg, share network'). For Coordination, optimal performance is more likely to arise when networks are distributed, negotiated or split. For Information-flow, optimal performance is likely to involve a Split of the command structure (2. Split and 6 Split, share). According to the SAS-050 definition, both structures are 'Hierarchical' networks with 'peer-to-peer' allocation of decision rights (they differ on the distribution of information variable). The next structures are 5. Cent, share and 1. Cent. Again, both are Hierarchical networks.

Agent Models

Introduction

An emerging theme within the literature, which might be taken as suggestive of the direction that the dominant cybernetic paradigm is taking, embodies Morgan's (1986) metaphor of organisations as a brain; resilient and flexible, capable of rational and intelligent change (Arnold et al., 1995). The language used to manifest this shift in structural modelling is related to the language (literally) and metaphors associated with computing and agent models (covered briefly in this section), and those associated with cognitive architectures.

Example: Work Environment Analysis (WEA)

'Declarative approaches to modelling are noted for a direct and concise way of stating facts and constraints and are therefore suitable for specification of environmental elements' (Shah and Pritchett, 2005, p.70). Declarative approaches are supported by contemporary computer languages such as XML (extensible mark-up language), and, as an example, have been used in the context of a method called Work Environment Analysis (WEA; Shah and Pritchett, 2005). The core of the approach is that pertinent aspects of a command and control environment are computationally modelled as reconfigurable nodes, in which these nodes are linked to other nodes according to specified relations. Manipulations made using XML in the specification of context/ process nodes/links enable changes to the Work Environment to be modelled and for the effect of these on system assessment criteria such as safety, stability and efficiency to be made. This approach relates to the HEAT method above, where in essence the use of a computer language and metaphors enables greater sophistication

Agent-based modelling

Joslyn (1999) writes that programming theory (as understood in the realm of computer science) has progressed 'from procedural through functional to object orientated models (such as XML), now culminating in [...] agent-based approach[es]' (p.2). This progression is analogous at some level to the modelling approaches in C4i. For example, the cybernetic approach might be considered as being at a procedural level, Lawson's and similar models are at a functional level, WEA (above) is object orientated, whereas this section is concerned with agent based approaches.

Agent-based models have two essential components: a representation of a set of active entities (actors) and a representation of an environment. Agents programmed with a range of behaviours on the basis of some form of rule or state system then interact with other agents and their environment to produce behaviour. One can then use this 'test bed' to examine the effects either of changes to the environment or changes to the rules which agents operate under (which might, for example, simulate a new technology or SOP). The qualities of the environment and the agent behaviour rules then constitute the key constraints within agent-based models.

The key advantage in using agent models is that as a product of the interaction between agents, emergent behaviour is produced. Emergence is a philosophically sophisticated notion that first became popular around the end of the 19th century. It was defined by John Stuart Mill in *A System of Logic* (1843):

> All organised bodies are composed of parts, similar to those composing inorganic nature, and which have even themselves existed in an inorganic state; but the phenomena of life, which result from the juxtaposition of those parts in a certain manner, bear no analogy to any of the effects which would be produced by the action of the component substances considered as mere physical agents. To whatever degree we might imagine our knowledge of the properties of the several ingredients of a living body to be extended and perfected, it is certain that no mere summing up of the separate actions of those elements will ever amount to the action of the living body itself.

It is important to note the irreducible nature of emergent properties, by definition one can observe them but not trace them back to any one element of the system that produced them. No individual part of the system has the emergent quality observable within it; emergent properties are a product of synergy. As a result of this irreducibility emergent properties will tend to be nonlinear in their response to changes within their underlying variables. If one takes the philosophical position that certain forms of behaviour are observable only as a result of complex interactions (and thus are non-linear in their response to changes in parameters, effectively putting it outside conventional analytical techniques), then agent modelling is arguably the only means by which emergent properties of systems can be explored. C4i can be considered from an agent based modelling perspective.

Semiotic agents

The active entities (the agents themselves) in an agent based modelling paradigm can be characterised according to the following five properties (Joslyn, 1999).

- *Asynchronicity*: Agents act independently in time.
- *Interactivity*: Agents communicate and interact in something analogous to a 'social' manner, forming a collective entity through their interaction.
- *Mobility*: Agents have the capacity for 'movement', in terms not just of movement through real, virtual or simulated space, but also notions of movement of data within or between environments.
- *Distribution*: Which is the manifest property of Mobility; and
- *Non-Determinism*: In which agent systems possess some aspects that can be regarded as random (Joslyn, 1999).

The term 'semiotic agent' refers to 'the use and communication of symbols by and between agents and their environments' (Joslyn, 1999, p.2). 'They involve processes of perception, interpretation, decision, and action with their environments' (Joslyn, 1999, p.9) and are therefore analogous to some extent with a three stage model of human information processing (for example, Newell and Simon, 1972). Perhaps because of this, semiotic agents have a certain appeal when it comes to modelling the role of human agents in systems. Semiotic agents possess the following architecture (Figure 2.16).

Example: BOIDS

An early finding within the agent modelling community was that complex group behaviour need not be a function of complex individual behaviour; even agents following very simple rules will produce complex patterns of interaction. Probably the most famous example of this synergistic complexity effect is to be found in the model of bird flocking behaviour 'Boids' (Reynolds, 1987). The aggregate motion of a flock of birds or other animals constitutes a deeply complex pattern of motion to describe in mathematical terms (as might be attempted above under a cybernetic paradigm). Reynolds (1987) found that if he modelled each individual 'boid' as an entity that acted according to a simple set of rules, but otherwise independently of other 'boids', a highly realistic flocking motion path was produced as an emergent property at the level of the group arising from interactions between individual entities. This occurred purely through individual boids following three rules: (i) a boid will avoid collisions by adjusting course; (ii) boids will attempt to match velocity with nearby boids; and (iii) boids will attempt to stay close to other nearby boids. Fidelity to realistic behaviour was found even when objects were introduced into the environment that the 'boids' had to navigate around (see Figure 2.17).

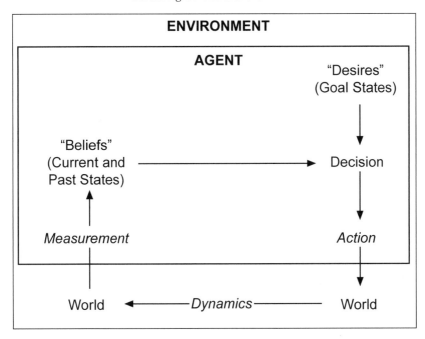

Figure 2.16 Architecture of a semiotic agent

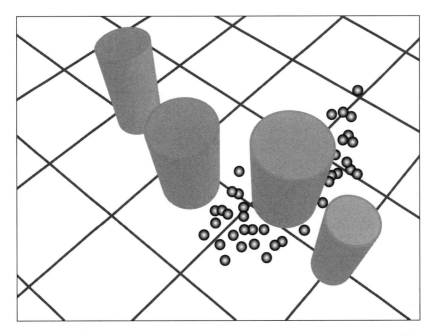

Figure 2.17 Simulated agents produce emergent flocking behaviour within a computer simulation

Summary of agent-based approaches

As regards modelling any complex system then, agent modelling provides the possibility of an alternate form of attack if traditional mathematical techniques (or the analyst's deployment of them) are insufficient to represent the complex, emergent and possibly non-linear behaviour of a system. One may raise the objection, however, that human behaviour is more intelligent than that of boids or indeed birds, but militating against this view is long standing evidence that significant portions of human behaviour may indeed be predicted by things like simple Pavlovian associative learning rules (Rescorla and Wagner, 1972). There may indeed by facets of a C4i system, such as procedures and rules that militate behaviour further.

Paradoxically, the agent modelling focus is still couched within the cybernetic paradigm, as the behaviour of nodes is still represented mathematically to some extent. The critical difference is that the underlying (simple) control model in this case is held within individual agent nodes, and the interactions among agents endow such models with (complex) aggregate behaviour. This at least gives the impression of much less formal determinism and can help to circumvent highly complex modelling mathematics.

Socio-technical Models

People and technology

Socio-technical systems, simply put, are a mixture of people and technology. As a modelling entity it represents '[…] an approach to complex organisational work design that recognises the interaction between people and technology in workplaces' (Wikipedia, 2005). Command and control scenarios are a particularly complex example of these systems. Structural, deterministic approaches to modelling confer a metaphor of 'organisations as machines' (Morgan, 1986). As such, human agents tend to be modelled around the structure, rather than the other way around. An alternate metaphor, command and control as an organism, appears to capture the essence of the socio-technical perspective rather better. Embodied within this type are the characteristics of adaptability and dynamism, which lends itself well to an environment that is more uncertain and turbulent. The challenge of course is to model such situations. An adaptation of Kotter's (1978) model of organisational dynamics presents the range of interacting, typically non-linear and overlapping facets that comprise a typical socio-technical system (Figure 2.18). This model is merely descriptive.

There are a number of different ways to visualise or otherwise represent the facets illustrated in Figure 2.18.

Example: Process charts

Process charts offer a systematic approach to describing activities within socio-technical systems. They simplify complexity by emphasising key features using a graphical representation that is easy to follow and understand (Kirwan and Ainsworth,

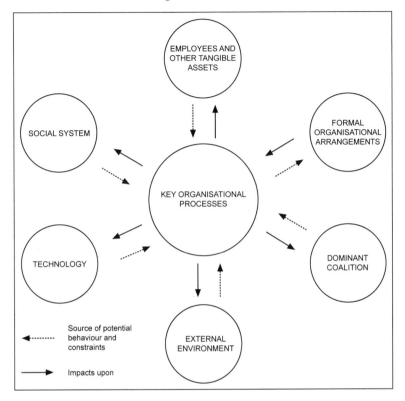

Figure 2.18 Kotter's model of organisational dynamics (1978)

1992). Process charts are descriptive in that they represent 'activity' over time, who (and/or what) is enacting any given activity and how activities are linked in order to culminate in the final goal based objective(s).

Process charts also offer some deeper predictive insights. Because activities are mapped along a timeline, outcome measures of tempo (number of activities within timeframes) can be gained. Operations loading (the type of activity versus who is performing the activity) provide some, albeit crude indication of workload. Although used in a predominantly descriptive manner there seems little reason why process charts could not be constructed before a task has taken place to form a simple type of predictive model. The construction, at a superficial level, could proceed according to a task analysis (itself a form of model), the times associated with the completion of specified tasks (a Keystroke Level Model) and any known mediating affects of communications media (expected errors etc.). Fundamentally, whilst process models of socio technical systems are adept at describing the observable outputs (and inputs) of human information processing, alternate approaches are required to model the 'unobservable' processes of human cognition; which are the main cause of modelling difficulties under more structural perspectives.

Example: A functional model of command and control

Smalley (2003; Figure 2.19) proposes a functional socio-technical model of command and control, comprising some seven operational and decision support functions (six in the ovals and one in the box) and ten information processing activities (appended to the input and output arrows). These include several 'unobservable processes'. The ten information processing activities are: primary situation awareness, planning, information exchange, tactical situation reports, current situation awareness, directing plan of execution, system operation, system monitoring, system status, and internal co-ordination and communications. A representation of the relationship between the operation and decision support functions and information processing activities is shown in Figure 2.19.

Information about the state of the world is collected through the primary situation awareness activities. The various sources of information are combined so that targets and routes can be defined in the planning activities. Information about targets, routes and intentions is exchanged with other forces. The status of the mission is communicated through the tactical situation reports. Current situation awareness activities merge information about the mission with primary situation awareness, to inform the planning process. The information from this latter set of activities will cue the start of activities that direct the plan of execution. This, in turn, informs activities associated with the direction of system operation. The system is monitored, to see if outcomes are as expected. Any changes in the system status may lead to changes in

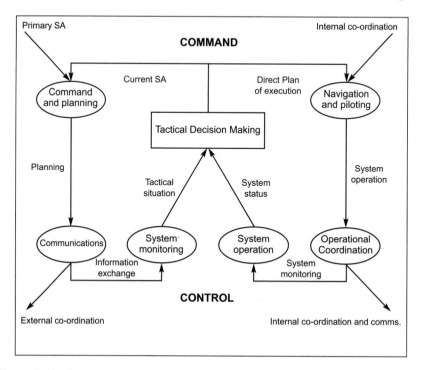

Figure 2.19 Smalley's functional command and control model

the planning and the directing of the plan. Internal and external co-ordination and communication activities keep the command and control system functioning.

Smalley's model seems to represent an integration of many command and control activities coupled together with feed-forward and feed-back loops. It has a higher level of command and control fidelity than previous model examples. The model suggests that 'command' activities (at the top of the figure) are separate, but connected to, the 'control' activities (at the bottom of the figure). The activities on the right-hand side of the figure are concerned with internal operation of the system, whereas the activities on the left-hand side of the figure are concerned with interfacing with the external environment.

Example: Cognitive Work Analysis (CWA)

The term cognitive engineering was first coined by, amongst others, Donald Norman, Erik Hollnagel and David Woods in relation to an expanding field of applied cognitive science based on the fundamental principle of human action and performance. Numerous cognitive engineering approaches fall under the umbrella of Cognitive Work Analysis (CWA; Vincente, 1999), which is to be discussed below as a paradigm. CWA attempts to offer predictive power in terms of the relationships specified in Kotter's diagram (Figure 2.18) with a focus on the goal directed behaviour of agents within the system. CWA involves five modelling phases as follows:

1. Work domain analysis,
2. Activity analysis,
3. Strategies analysis,
4. Socio-organisational analysis, and
5. Worker competencies analysis.

Work domain analysis
The work domain analysis involves describing the work environment using an abstraction decomposition space. The principal behind this lies in the premise that 'instead of decomposing functions according to the structural elements, we have to abstract from these elements and [...] identify and [...] separate the relevant functional relations' (Rasmussen et al., 1994). In the Abstraction Decomposition Space (ADS) there are five levels of abstraction, ranging from the most abstract level of purposes (for example, the intended effect of the whole system upon the environment) to the most concrete level of form (for example, the physical form of actual objects in the environment; Vicente, 1999). Table 2.5 presents a definition of the cells within the abstraction decomposition space and presents a worked example from a militaristic work domain. Moving up or down the ADS varies the level of abstraction, whereas moving left to right is the equivalent of zooming in or out on the system. When levels of abstraction are crossed with levels of system description an explanatory model of the 'reasons that a technical system exists and must be controlled' is gained (Sanderson, 2003). 'Connections between an element at one level to an element at the next level above indicates why the first element exists,

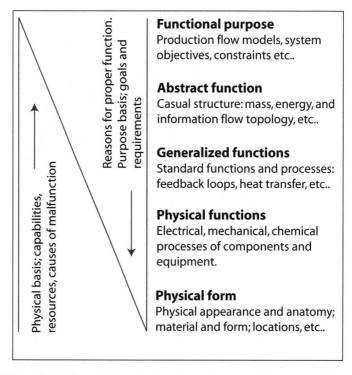

Figure 2.20 Definition of the abstraction decomposition space (Rasmussen, 1986)

Table 2.5 Worked example of an abstraction decomposition space from a militaristic work domain

AH/SH	Total system	Sub-system	Functional Unit	Assembly	Component
Functional purpose	Overall mission	Command level goals	Unit goals	Team goals	Agents goals
Abstract function	Mission plans	Mission plans	Tactical overlays	Tactical overlays	Projected agent
Generalised function	Course of action	Sub-system capability	Unit capability	Team capability	Agent capability
Physical function	Mission status	Mission summaries	Unit mission summaries	Team status	Agent status
Physical form	Global view of BS	Location of sub-system	Location of unit	Location of team	Location of agents

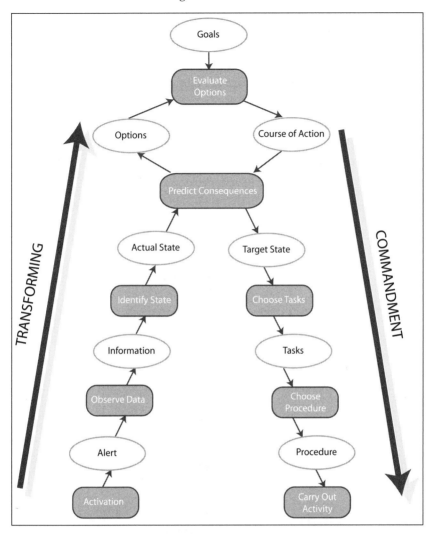

Figure 2.21 Rasmussen's decision-ladder applied to a command and control domain (Chin et al., 1999)

whereas connections to an element at the next level below indicates how the element has been instantiated or engineered' (Sanderson, 2003, p.241).

Activity analysis
The activity analysis component involves identifying the tasks that need to be performed in the work domain. Various task analysis techniques relate to this step of CWA and the reader is referred to the section on Hierarchical Task Analysis (HTA) above as an example. Regardless of the specific modelling technique the aim is the same; to '[…] identify what needs to be done, independently of how or by whom, using a

constraint based approach' (Vicente, 1999, p.183). Information processing constraints placed upon the enactment of tasks are typically modelled using the decision ladder from Rasmussen et al. (1994), reproduced in relation to command and control tasks in Figure 2.21. 'The decision ladder is not a model of human decision making but instead is a template of possible information-processing steps that allow a controller [human or non-human] to take information about the current state of the system and execute appropriate actions in response' (Sanderson, 2003, p.243).

Strategies analysis
Strategies analysis involves modelling the mental strategies that the agents involved may use during task performance in the domain under analysis. But, 'rather than descriptions of the course and content of actual mental processes [which is hard if not impossible to accurately model], descriptions of the structure of possible and effective mental processes' are represented (Rasmussen, 1981, p.242). This, arguably,

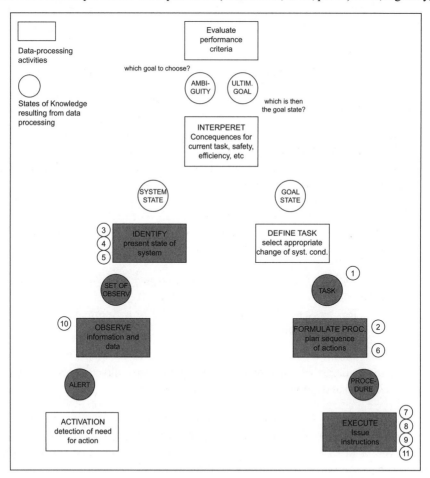

Figure 2.22 Example output of strategies analysis. Shaded regions refer to a specific role or actor; the numbered circles refer to task steps

is CWA's least defined representation. Typically, however, strategies analysis takes the form of flow charts (see Figure 2.22).

Social organisation analysis
Social organisation analysis involves identifying exactly how the work is distributed amongst the agents and artefacts within the system under analysis. In CWA this analysis usually proceeds by overlaying some of the previous representations. It would appear that the network techniques described are also highly relevant to modelling this aspect of coordination between actors.

Worker competencies analysis
Finally, worker competencies analysis involves the identification of the competencies that the agents involved are required to possess in order to perform the task(s) in the work domain. The worker competencies are classified using Rasmussen's Skill, Rule, Knowledge (SRK) framework. The SRK framework is a 'taxonomy for classifying how cognition is controlled by the way information is presented in the environment' (Sanderson, 2003, p.250). As such its modelling attributes are more concerned with simplifying complexity, perhaps pointing the way to further distributed models of human cognition.

In terms of outcome measures, CWA provides various perspectives on purpose, intent, goals and capability (Chin et al., 1999), as well as to the attribution of information, systems and activities to tasks and timescales.

Example: Contextual Control Model

Whereas CWA is an integrative approach to modelling socio technical systems, Hollnagel's (1993) Contextual Control model is a contemporary example of how human performance within such systems might be represented. Hollnagel (1993) developed a Contextual Control approach to human behaviour, based on cognitive modes to explain the effects of the context in which people performed their actions. Rather than command and control being a pre-determined sequence of events, Hollnagel has argued that it is a constructive operation in which the operator actively decides which action to take according to the context of the situation together with their own level of competence. Although set patterns of behaviour may be observed, Hollnagel points out that this is reflective of the environment as well as the cognitive goal of the person, both of which contain variability. In the Contextual Control Model (shown in Figure 2.23), four proposed modes of control are offered as follows.

- *Scrambled Control:* is characterised by a completely unpredictable situation in which the operator has no control and has to act in an unplanned manner, as a matter of urgency.
- *Opportunistic Control:* is characterised by a chance action taken due to time, constraints and again lack of knowledge or expertise and an abnormal environmental state.

- *Tactical Control:* is more characteristic of a pre-planned action, where the operator will use known rules and procedures to plan and carry out short term actions.
- *Strategic Control:* is defined as the 'global view', where the operator concentrates on long term planning and higher level goals.

All these modes of control are familiar within C4i scenarios.

The degree of control is determined by a number of varying interdependent factors. Hollnagel considers that availability of subjective time is a main function of command and control – this means that as the operator perceives more time available so they gain more control of the task/situation. In fact one could argue that the fundamental

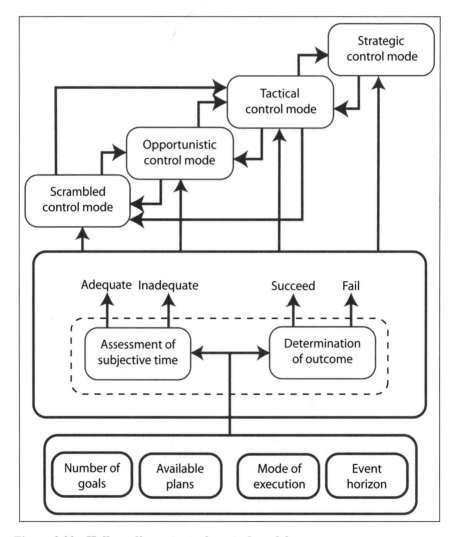

Figure 2.23 Hollnagel's contextual control model

purpose of C4i is to facilitate increases in subjectively available time and by doing so increasing 'decision superiority'; striving to operate at strategic/tactical levels of control. The factors affecting the perception of available time may include: the number of goals, the availability of plans to meet these goals, the modes of execution of those plans and the time available before (called the event horizon). At any point in time the system operator is attempting to optimise all of these criteria. Hollnagel's model differs from Lawson's in that it does not prescribe the sequence and relations of command and control activities; rather it proposes contextual differences in the control mode. There is some evidence to support this hypothesis. In a study of team behaviour in a supervisory control task, Stanton et al. (2001) showed that the transitions between control modes were consistent with Hollnagel's model.

Summary of Socio-technical Models

If socio-technical systems are regarded merely as mixtures of people and technology then the models reviewed in this section aim to characterise and predict the behaviour of people in context. Vincente's (1999) pioneering CWA method is a comprehensive integrative approach that places a particular emphasis on modelling the context from a cognitive perspective. What might be regarded as cognitive engineering models, for example, Rasmussen's and Hollnagel's approaches, are 'general characterisations' of human behaviour in systems and in environments. Hollnagel's model places emphasis on subjectively available time and links it to four levels of 'control'. Rasmussen's model (as it is embodied in CWA) characterises naturalistic decision making, expressing shortcuts and routes through a hierarchy of decision making. Functional models, in essence, break down the mixture of people and technology into purposeful elements with links and feedback. Process charts characterise socio-technical systems in terms of activity.

Summary of Modelling Review

This chapter defines four broad modelling typologies: Cybernetic, Network, Agent Based and Socio-Technical.

The cybernetic modelling paradigm is concerned principally with the structural aspects of command and control, reducing it to functional entities linked through specific causal pathways according to a deterministic idiom. These models can be subject to various known inputs and the specification of the functional entities enables the resulting output to be completely described.

Network models blur somewhat the strict formalism of the cybernetic perspective. The focus widens to emphasise not just the functional entities themselves but also the links that exist between them. The links can be defined according to various parameters, including communications between functional elements and logical relationships. When functional entities are linked in this manner a network is formed. The network rather than the functions can be summarised and analysed mathematically to reveal emergent properties. The emergent properties are not necessarily planned *a priori* thus the network approach provides an alternate perspective on, as well as a prediction of several C4i system attributes and outcomes.

Agent modelling perspectives appear to represent a form of synthesis between cybernetics and network models. Whereas cybernetic models attempt to model the 'aggregate behaviour' of a group of entities, doing so with often complex mathematics, agent approaches focus on the emergent behaviour arising from the interaction of (mathematically and computationally) simplistic entities. That is, complex group behaviour need not be a function of complex individual behaviour. Agent modelling results in less formal and more organic behaviour from which complex emergent properties arise.

Socio-technical models of command and control emphasise the human role. Rather than the strict formalism of the previous approaches cognitive models tend to be a more general characterisation of agent behaviour (and psychology) in C4i systems. Socio-technical systems, being a mixture of people and artefacts, aim to specify the environmental factors that influence human cognition and which form model constraints. Effective decision making and behaviour can be assumed to be the key emergent property, in which the interest is couched within the key determinates of C4i scenarios that facilitate or indeed hinder this outcome.

Table 2.6 presents a summary of the four broad modelling perspectives identified. The summary is represented in terms of emergent properties.

This chapter has already alluded to the fact that the possibilities and challenges underlying future NEC approaches are a primary motivation for a generic model of command and control. The expected outcomes of NEC are expressed as a benefits chain (MoD, 2004) shown in Figure 2.24.

Whilst this chapter is not intended to be an exhaustive critique of individual models, it is however informative to relate the examples given above to the benefits chain as shown in Table 2.7.

Table 2.7 represents a relatively informal classification. The cybernetic models can be associated with modelling 'effects', these can be regarded as the more

Table 2.6   Summary of modelling perspectives and broad emergent properties

Modelling perspective	Emergent properties
Cybernetics	Structural C4i system parameters (such as casualties, survivability, mission success etc) in relation to known inputs.
Networks	Characteristics of networks that enhance or constrain performance.
Agents	Aggregate behaviour based on the interaction between entities.
Cognitive/STS Models	Constraints on effective human decision making and action.

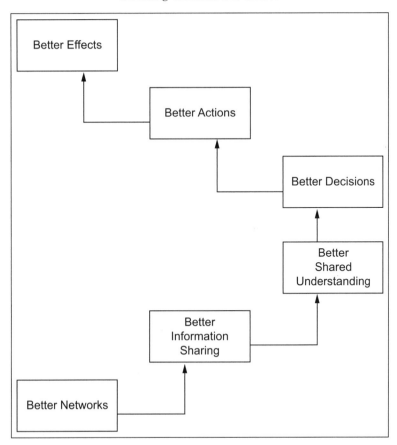

Figure 2.24 NEC benefits chain

structural parameters of mission success (such as destroy, find etc.). The benefits related to the network models depend on the 'topic' of the network. Networks that possess links related to information and communication can be loosely associated with benefits of information sharing. Networks expressing the links between tasks and goals relate to 'NEC Networks', that is the physical instantiation of a real life NEC system. Propositional networks, related as they are to knowledge clearly speak towards benefits in shared understanding. Agent models all seem to relate to actions, being concerned with modelling 'aggregate' behaviour. Similarly, cognitive and STS models relate to cognition in context as it relates to the underlying determinates of human decision making and action in C4i systems.

Whilst individually each of the models and model types speak towards different aspects of C4i and NEC, none of them can be considered a 'generic model' in itself. It can be argued that C4i is too multi-faceted for the modelling typologies dealt with so far to encompass the full range of dimensions and emergent behaviour relevant for military research. An intermediate modelling step is therefore required in order to:

Table 2.7 Informal classification of model types/typologies with defined NEC benefit criteria

Modelling perspective	Example model	Related benefits
Cybernetic	Lawson's Model of C2	Effects
	HEAT	Effects
Network	Hitchen's N2	Information Sharing
	HTA	Networks
	Social/Prop Nets	Information Sharing/Shared Understanding
Agent	WEA	Actions
	BOIDS	Actions
Cognitive/STS models	Process Model	Actions
	Functional Model	Decisions/Actions
	CWA	Decisions/Actions
	COCOM	Decisions/Actions

- simplify complexity;
- enable emergent properties and outcome metrics to be defined;
- enable some of the modelling typologies identified above to be selected from and deployed intelligently in future research;
- enable C4i scenarios from alternate domains and contexts to be modelled on a common platform; and
- for these disparate scenarios to be easily compared.

The manifestation of these aims is presented in the rest of this book. A methodology for analysing complex command environments is presented in the next chapters. Examples of the application of the methodology to three different domains are presented in the following three chapters. In the final chapters a generic process model of C4i is developed.

Chapter 3

Event Analysis of Systemic Team-work

With contributions from Dan Jenkins, Paul S. Salmon and Guy H. Walker

EAST Review

The event analysis of systemic team-work (EAST) methodology (Baber and Stanton, 2004; Stanton et al., 2005) was developed as part of the HFI DTC work programme, and uses a combination of HF methods to form a framework for analysing command, control, communication, computers and intelligence (C4i) activity. With a view to the development of a generic model of C4i, EAST has been applied to a number of diverse C4i scenarios, across a number of different domains. The domains and scenarios in which EAST has been used to analyse C4i activity are presented in Table 3.1.

In order to analyse the performance of the EAST methodology and its component methods, a review of the method was conducted based upon the applications described above. The review was based upon the same criteria that were used in the HF methods review (Salmon et al., 2004a) that was conducted for the development of the EAST methodology. The criteria are outlined below:

1. Name and acronym – the name of the technique and its associated acronym.
2. Author(s), affiliations(s) and address(es) – the names, affiliations and addresses of the authors are provided to assist with citation and requesting any further help in using the technique.
3. Background and applications – this section introduces the method, its origins and development, the domain of application of the method and also application areas that it has been used in.
4. Domain of application – describes the domain that the technique was originally developed for and applied in.
5. Procedure and advice – this section describes the procedure for applying the method as well as general points of expert advice.
6. Advantages – lists the advantages associated with using the method in the design of C4 systems.
7. Disadvantages – lists the disadvantages associated with using the method in the design of C4 systems.
8. Example – an example, or examples, of the application of the method are provided to show the methods output.
9. Related methods – any closely related methods are listed, including contributory and similar methods.
10. Approximate training and application times – estimates of the training and application times are provided to give the reader an idea of the commitment

Table 3.1 EAST analyses

Domain	Scenario
Army Battle Group Headquarters	Combat estimate
	Seven questions
Air Traffic Control	Overflight
	Departure
	Approach
	Shift handover
Energy Distribution	Barking switching operations
	Feckenham switching operations
	Tottenham return to service operations
	Alarm handling operations
Fire Service	Chemical incident at remote farmhouse
	Road traffic accident involving chemical tanker
	Factory fire exercise #1
	Factory fire exercise #2
Military Aviation E3D	General operation
Navy HMS Dryad	Air threat
	Surface threat
	Sub-surface threat
Police	Car break-in caught on CCTV
	Suspected car break-in
	Mobile phone robbery
Rail (Signalling)	Detachment scenario
	Emergency Possession scenario
	Handback a Possession scenario
	Possession scenario

required when using the technique. Training and application times are classified as low (0 – 2 hours), medium (2 – 6 hours) or high (6 + hours).
11. Reliability and validity – any evidence on the reliability or validity of the method are cited.
12. Tools needed – describes any additional tools required when using the method.
13. Bibliography – a bibliography lists recommended further reading on the method and the surrounding topic area.
14. Flowchart – a flowchart is provided, depicting the methods procedure.

The criteria were applied to the overall EAST methodology and also the component individual methods. A summary of the EAST review is presented in Table 3.2.

Methods Review

EAST – Event Analysis of Systemic Teamwork

Background and applications
The EAST methodology (Baber and Stanton, 2004) was developed for the analysis of C4i (command, control, communications, computers and intelligence) activity. EAST uses a combination of HF methods to form a framework for analysing C4i activity. A brief description of each component method is provided below:

- Hierarchical Task Analysis (HTA) – involves describing the scenario under analysis using a hierarchy of goals, sub-goals and operations.
- Observation – is used during an EAST analysis to gather data surrounding the scenario under analysis. The observational data obtained is used as the primary input to an EAST analysis.
- Co-ordination demands analysis (CDA) – involves defining the task-work and team-work tasks involved during the scenario, and then rating each teamwork task step against the CDA taxonomy of communication, situational awareness, decision making, mission analysis, leadership, adaptability and assertiveness.
- Comms Usage Diagram (CUD) – is used to describe collaborative activity between distributed agents. The output of CUD describes how and why communications between a team occur, which technology is involved in the communication, and the advantages and disadvantages of the technology medium used.
- Social Network Analysis – involves defining and analysing the relationships between agents within a scenario network. A matrix of association and a social network diagram are constructed, and agent centrality, sociometric status and network density figures are calculated.
- Operation Sequence Diagram (OSD) – is used to represents the tasks, the actors, the communications, the social organisation, the sequence and time in which the scenario took place. An OSD captures the flow of information

Table 3.2   Summary of EAST methods review

Method	Type of method	Domain	SMEs	Related methods	Training time	Application time	Tools needed
Event Analysis of Systemic Teamwork (EAST) (Baber and Stanton, 2004)	Team analysis method	Generic C4i	Yes	Obs HTA CDA OSD CUD SNA CDM Prop Nets	High	High	Video / Audio recording equipment MS Word MS Excel MS Visio AGNA
Observation	Data collection	Generic	No	HTA	Low	High	Video / Audio recording equipment MS Word
Hierarchical Task Analysis (Annett, 2004)	Task analysis	Generic	No	Obs	Med	High	Pen and paper MS Notepad
Co-ordination demands analysis (CDA) (Burke, 2005)	Team analysis	Generic	Yes	Obs HTA	Low	Med	Pen and paper MS Excel

Reliability	Validity	Advantages	Disadvantages
Med	High	1. EAST offers an extremely exhaustive analysis of the C4i domain in question. 2. EAST is relatively easy to train and apply. The provision of the WESTT and AGNA software packages also reduces application time considerably. 3. The EAST output is extremely useful, offering a number of different analyses and perspectives on the C4i activity in question.	1. Due to its exhaustive nature, EAST is time consuming to apply. 2. Reliability may be questionable in some areas. Reliability of the method is still being established. 3. A large portion of the output is descriptive. Great onus is placed upon the analyst to interpret the results accordingly.
High	High	1. Acts as the primary input for the EAST methodology. 2. Easy to conduct provided the appropriate planning has been made. 3. Allows the analysts to gain a deeper understanding of the domain and scenario under analysis.	1. Observations are typically time consuming to conduct (including the lengthy data analysis procedure). 2. It is often difficult to gain the required access to the establishment under analysis. 3. There is no guarantee that the required data will be obtained.
Med	High	1. Easy to learn and apply. 2. Allows the analysts to gain a deeper understanding of the domain and scenario under analysis. 3. Describes the task under analysis in terms of component task steps and operations.	1. Can be difficult and time consuming to conduct for large, complex scenarios. 2. Reliability is questionable. Different analysts may produce different HTA outputs for the same scenario. 3. Provides mainly descriptive rather than analytical information. Also does not cater for the cognitive components of task performance.
Low	High	1. Offers an overall rating of co-ordination between team members for each teamwork based task step in the scenario under analysis. 2. Also offers a rating for each teamwork behaviour in the CDA teamwork taxonomy. 3. The technique can be used to identify task-work (individual) and team-work task steps involved in the scenario in question.	1. The CDA procedure can be time consuming and laborious. For each individual task step, seven teamwork behaviours are rated. 2. To ensure validity, SMEs are required. 3. Intra and inter-analyst reliability may be questionable.

among distributed actors and shows how this is mediated through technology and team working. Additionally, operational loading figures are calculated for the following OSD operators: Operation, Receive, Delay, Decision, Transport, Combined Operations.
- Critical Decision Method (CDM) – involves the use of interview probes to elicit information regarding agent decision making strategies adopted during the scenario under analysis.
- Propositional Networks – The CDM output is used to construct propositional networks for each incident phase. Propositional networks are comprised of nodes that represent sources of information, agents, and objects that are linked through specific causal paths. The propositional network thus represents the 'ideal' collection of knowledge for an incident.

The EAST methodology has been applied in a number of domains, including the fire service (Baber et al., 2004a), police service (Baber et al., 2004b) naval warfare (Stewart et al., 2004), military aviation, energy distribution (Salmon et al., 2004b; 2004c), air traffic control and rail domains (Walker et al., 2004), in order to analyse C4i activity. EAST is a comprehensive technique offering a multi-faceted assessment of the C4i network in question. EAST provides an assessment of agent roles within the network, a description of the activity including the flow of information, the component tasks, communication between agents and the operational loading of each agent. Co-ordination between agents is also rated and the knowledge required throughout the task under analysis is defined.

Domain of application

Generic. The EAST methodology can be used in any domain that utilises a C4i infrastructure during task performance.

Procedure and advice

Step 1: Define scenario(s) The first step in an EAST analysis involves defining the scenario(s) that are to be the subject of the analysis. It is recommended that this is done in collaboration with an SME from the domain in question. The chosen scenario(s) should be representative of all C4i activity in the domain in question. For example, in an analysis of C4i activity in the civil energy distribution (Salmon et al., 2004b; 2004c), a switching out of circuit's scenario and a return to service of circuits scenario were used.

Step 2: Conduct HTA for the scenario(s) under analysis Once they are clearly defined, the analyst(s) should conduct a HTA for each scenario. Again, it is recommended that this is done in collaboration with a relevant SME.

Step 3: Plan observation The next step of the EAST analysis involves planning the observation of the scenario(s) under analysis. Typically the scenarios involved in C4i activity are dispersed over a number of different locations involving a number of agents, and so who to observe and where to observe it requires careful thought. Again, input from an appropriate SME is useful here. It is also useful at this stage to clearly define what data are to be collected. The use of recording equipment (for example, type, location, focus etc.) should also be clarified. If time permits, it may also be pertinent to conduct a trial run of the planned observation.

Step 4: Observe scenario(s) The observation step is the most important part of the EAST procedure. Typically, a number of analyst(s) are used in scenario observation. All activity involved in the scenario under analysis should be recorded along an incident timeline, including a description of the activity, the agents involved, any communications made and the technology involved. Additional notes should be made where required, including the purpose of the activity, any errors made and also any information that the agent involved feels is relevant.

Step 5: Critical decision method Once the scenario under analysis is complete, each 'key member of personnel' (for example, scenario commander, agents performing the component tasks etc.) involved should be subjected to a CDM interview. This involves dividing the scenario into key incident phases and then interviewing the agent involved using pre-defined probes. A CDM interview should be conducted for each incident phase.

Step 6: Transcribe scenario data Once all of the scenario data are collected, it should be transcribed in order to make it compatible to the EAST analysis phase. An event transcript should be constructed. The transcript should describe the scenario over a timeline, including descriptions of activity, the agents involved, any communications made and the technology used. In order to ensure the validity of the data, the scenario transcript should be reviewed by one of the SMEs involved.

Step 7: Re-iterate HTA The data transcription process allows the analyst(s) to gain a deeper and more accurate understanding of the scenario under analysis. It also allows any discrepancies between the initial HTA scenario description and the actual activity observed to be resolved. Typically, C4i activity does not run entirely according to protocol, and certain tasks may have been performed during the scenario that were not described in the initial HTA description. The analyst should compare the scenario transcript to the initial HTA, and add any changes as required.

Step 8: Co-ordination demands analysis The CDA involves extracting teamwork tasks from the HTA and rating them against the associated CDA taxonomy. Each teamwork task is rated against each CDA behaviour on a scale of 1 (low) to 3 (high). Total co-ordination for each teamwork step can be derived by calculating the mean across the CDA behaviours. The mean total co-ordination figure for the scenario under analysis should also be calculated.

Step 9: Construct comms usage diagram The CUD is used to represent communication between the agents involved in the scenario and also to analyse the communications technology utilised. The scenario transcript is used as the input to the CUD.

Step 10: Conduct social network analysis   The SNA is used to analyse the relationships between the agents involved in the scenario under analysis. Using the scenario transcript, the analyst should firstly construct an agent association matrix, which presents the links between the agents and also the frequency of communications between the agents. An example association matrix is presented in Table 3.6. From the association matrix, a social network diagram is constructed and agent centrality, sociometric status, and network density are calculated. It is recommended that the AGNA SNA software package is used for the SNA phase of the EAST methodology.

Step 11: Construct operation sequence diagram   The OSD represents the activity observed during the scenario under analysis. The analyst should construct the OSD using the scenario transcript and the associated HTA as inputs. Once the initial OSD is completed, the analyst should then add the results of the CDA to each teamwork task step.

Step 12: Construct propositional networks   The final step of the EAST analysis involves constructing propositional networks for each scenario phase identified during the CDM interviews. In order to construct the propositional networks for each phase, a basic contents analysis should be conducted using the associated CDM outputs. Knowledge objects are defined as knowledge, information, artefacts and actions. Each knowledge object should have a corresponding node in the propositional network. Next, the links between the nodes should be specified. To do this, the analyst should specify any links between the knowledge objects within the propositional networks, using the following links taxonomy:

- has
- is
- causes
- knows
- requires
- prevents.

Advantages

- The EAST approach offers an exhaustive analysis of C4i activity. Activity is analysed using a number of different approaches, ensuring that a comprehensive analysis of the scenario under analysis is achieved.
- The EAST output is extremely useful, offering a number of different analyses and perspectives on the C4i activity in question. Each component method offers a different representation or analysis of the C4i activity observed. For example, the SNA provides an indication of the key agents within the scenario and identifies communication links, the CDA identifies teamwork tasks and provides a co-ordination rating for each teamwork task step, and the propositional network component identifies the ideal knowledge (that is, SA) required for effective task performance.

- The methods within the EAST framework are conceptually simple, ensuring that the methodology requires minimal training and is relatively easy to apply.
- The provision of WESTT and AGNA software packages reduces the EAST application time considerably.
- The EAST methodology is generic, allowing it to be applied in any domain where C4i activity takes place. To date, the EAST methodology has been used to analyse C4i activity in a number of diverse scenarios, including fire service training scenarios (Baber et al., 2004a), military aviation scenarios (Stewart et al, 2008), railway signalling tasks (Walker et al., 2006a), air traffic control scenarios (Walker et al., 2005), energy distribution scenarios (Salmon et al., 2004b; 2004c) and naval warfare scenarios (Stewart et al., 2008).
- The methodologies that make up EAST share a common theoretical background. This makes integration viable (and embodies a form of congruent validity).
- There are possibilities for further recombination of methods to explore issues outside of the EAST framework (Walker et al., 2005a).
- All the methods have a validation history and prior context of use.

Disadvantages

- Due to the technique's exhaustiveness it is time consuming to apply. The initial observation, construction of the HTA and OSD are particularly time consuming elements of the methodology.
- The data obtained during the CDM analysis is a function of the analyst's skill and the quality of the SME used.
- Reliability of the technique in some areas is questionable. Much of the analysis is based upon the subjective judgement of the analyst involved, and so intra-analyst and inter analyst reliability may suffer.
- For the technique to work properly, access to the system and personnel under analysis is required. It is often difficult to gain sufficient access to the system under analysis. Further, additional access to SMEs for data validation purposes is also difficult to obtain.
- The methodology is still evolving and, as a consequence, some EAST outputs may differ in format and presentation.
- Without the provision of the appropriate software, an EAST analysis may become laborious to perform for larger, complex scenarios.
- Some prior knowledge of HF methods is required.

Example

The following example is taken from an analysis of a switching scenario drawn from the civil energy distribution domain (Salmon et al., 2004b; 2004c). The scenario took place at Barking 275Kv, 132Kv and 33Kv substations and the National Grid Transco (NGT) Network Operations Centre (NOC) in Warwick. The scenario under analysis involved the switching out of three circuits (SGT5 and SGT1A and 1B) at Barking 275Kv, 132Kv and 33Kv Substations. Circuit SGT5 was being switched out for the installation of a new transformer for the nearby channel tunnel rail link and SGT1A and 1B were being switched out for substation maintenance. The observation focussed upon four main agents involved in the scenario. A description of the agents involved is presented in Table 3.3.

Activity was watched by two observers. The first observer was situated at the Network operations centre (NOC) at National Grid Transco (NGT) house in Warwick, and observed the activity of the NOC operator. The second observer was situated at Barking 275Kv, 132Kv and 33Kv Substations, and observed the activity of the SAP and AP. The results of the CDA are presented in Table 3.4. An extract of the CDA is presented in Table 3.5.

The agent association matrix is presented in Table 3.6. From the association matrix, a social network diagram was constructed (Figure 3.1). Directional arrows represent the communications between agents, and the frequency of these communications is also presented. Centrality and sociometric status were also calculated for each agent involved in the scenario. The SNA results are presented in Table 3.7.

A CUD was constructed in order to analyse the communications made and the communications technology used. An extract of the CUD for the switching scenario is presented in Figure 3.2.

An OSD was constructed in order to represent the flow of activity during the switching scenario. The OSD glossary is presented in Figure 3.3. An extract of the OSD for the switching scenario is presented in Figure 3.4. From the OSD, operational

Table 3.3 Agents involved in switching scenario

Agent Title	Details
NOC	Control room operator based at the NOC
SAP	Senior Authorised Person (SAP) at Barking substations. Also working with authorised person (AP)
WOK	Control room operator based at Wokingham control room
REC	Control room operator based at the regional electricity company control room

Table 3.4 Switching scenario CDA results

Category	Result
Total task steps	314
Total taskwork	114 (36%)
Total teamwork	200 (64%)
Mean Total Co-ordination	1.57
Modal Total Co-ordination	1.00
Minimum Co-ordination	1.00
Maximum Co-ordination	2.14

Table 3.5 Extract of CDA analysis

Task Step	Agent	Step No.	Task Work	Team Work	Comm	SA	DM	MA	Lead	Ad	Ass	TOT CO-ORD Mode	TOT CO-ORD Mean
1.1.1	WOK control room operator	Use phone to contact NOC	1										
1.1.2	WOK control room operator / NOC control room operator	Exchange identities		1	3	3	1	1	1	1	1	1.00	1.57
1.1.3	WOK control room operator / NOC control room operator	Agree SSC documentation		1	3	3	3	1	1	1	1	1.00	1.86
1.1.4.1	NOC control room operator	Agree SSC with WOK		1	3	3	3	1	1	1	1	1.00	1.86
1.1.4.2	NOC control room operator	Agree time with WOK		1	3	3	3	1	1	1	1	1.00	1.86
1.1.5.1	NOC control room operator	Record details onto log sheet	1										
1.1.5.2	NOC control room operator	Enter details into WorkSafe	1										
1.2.1	NOC control room operator	Ask for isolators to be opened remotely		1	3	3	1	2	2	1	1	1.00	1.86
1.2.2	WOK control room operator	Perform remote isolation	1										
1.2.3	NOC control room operator	Check barking s/s screen	1										
1.2.4	WOK control room operator / NOC control room operator	End communications		1	3	1	1	1	1	1	1	1.00	1.29
1.3.1	NOC control room operator	Use phone to contact SAP at Barking	1										
1.3.2	NOC control room operator / SAP at Barking	Exchange identities		1	3	3	1	1	1	1	1	1.00	1.57

Table 3.6 Agent association matrix

	A	B	C	D
A	0	3	2	1
B	8	0	2	2
C	3	3	0	0
D	1	1	1	0

Table 3.7 SNA results

Agent	B-L Centrality	Sociometric Status
SAP at Barking	2.16	6.33
NOC operator	2.16	6.00
REC operator	1.85	2.00
WOK control operator	1.85	3.66

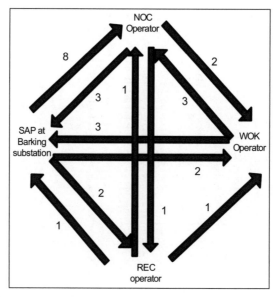

Figure 3.1 Social network diagram

Event Analysis of Systemic Team-work 61

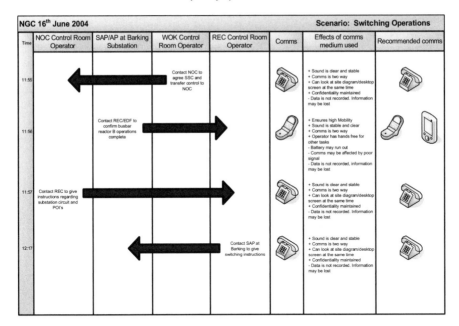

Figure 3.2   CUD extract

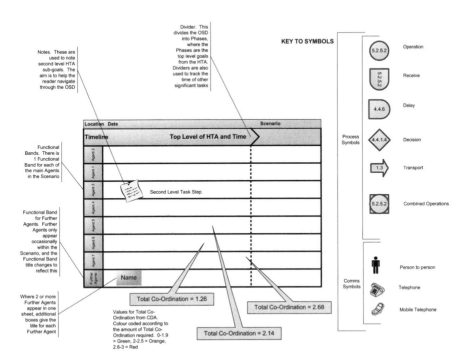

Figure 3.3   OSD glossary

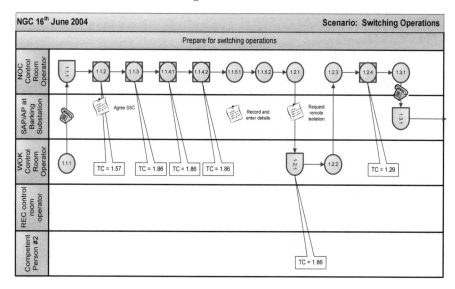

Figure 3.4 OSD extract

Table 3.8 Operational loading results

Agent	Operation	Receive	Transport	Decision	Delay	Total
NOC	98	40				138
SAP	223	21	19		1	264
WOK	40	10				50
REC	15	14				29

loading figures were calculated in order to determine agent loading during the scenario. The operational loading figures are presented in Table 3.8.

The operational loading figures indicate that the NOC operator was loaded the highest for receive (that is, receipt of information) operations, and the SAP at Barking was loaded the highest for operation, transport and delay operations.

The CDM analysis for the switching scenario involved interviewing the SAP at Barking and the NOC control room operator. In both cases the probes defined by O'Hare et al. (2000) were used to evaluate the decision making during the scenario. The CDM probes used are presented in Table 3.9. For the purposes of the CDM analysis, the scenario was divided into four phases; First issue of instructions, deal with switching requests, perform isolation and report back to NOC. The CDM analysis is presented in Tables 3.10 to 3.13.

Table 3.9   CDM probes (O'Hare et al., 2000)

Goal Specification	What were your specific goals at the various decision points?
Cue Identification	What features were you looking for when you formulated your decision? How did you know that you needed to make the decision? How did you know when to make the decision?
Expectancy	Were you expecting to make this sort of decision during the course of the event? Describe how this affected your decision making process.
Conceptual	Are there any situations in which your decision would have turned out differently? Describe the nature of these situations and the characteristics that would have changed the outcome of your decision.
Influence of uncertainty	At any stage, were you uncertain about either the reliability of the relevance of the information that you had available? At any stage, were you uncertain about the appropriateness of the decision?
Information integration	What was the most important piece of information that you used to formulate the decision?
Situation Awareness	What information did you have available to you at the time of the decision?
Situation Assessment	Did you use all of the information available to you when formulating the decision? Was there any additional information that you might have used to assist in the formulation of the decision?
Options	Were there any other alternatives available to you other than the decision you made?
Decision blocking - stress	Was there any stage during the decision making process in which you found it difficult to process and integrate the information available? Describe precisely the nature of the situation.
Basis of choice	Do you think that you could develop a rule, based on your experience, which could assist another person to make the same decision successfully? Why/Why not?
Analogy/ generalisation	Were you at any time, reminded of previous experiences in which a similar decision was made? Were you at any time, reminded of previous experiences in which a different decision was made?

Table 3.10 CDM Phase 1: First issue of instructions

Goal Specification	Establish what isolation the SAP at Barking is looking for. Depends on gear?
Cue Identification	Don't Believe It (DBI) alarm is unusual – faulty contact (not open or closed) questionable data from site checking rating of earth switches (may be not fully rated for circuit current – so additional earths may be required).
	Check that SAP is happy with instructions as not normal.
Expectancy	Decision expected by DBI is not common.
Conceptual Model	Recognised instruction but not stated in WE1000 – as there are not too many front and rear shutters metal clad switch gear.
Uncertainty	Confirm from field about planned instruction – make sure that SAP is happy with the instruction.
Information	Reference to front and rear busbars.
Situation Awareness	WE1000 procedure.
	Metal clad switchgear.
	Barking SGT1A/1B substation screen.
	SAP at Barking.
Situation Assessment	Ask colleagues if needed to.
Options	No alternatives.
Stress	N/A.
Choice	WE1000 – need to remove what does not apply.
	Could add front and rear busbar procedures.
Analogy	Best practice guide for metal clad EMS switching.

From the CDM outputs, propositional networks were constructed for each incident phase. The propositional networks are presented in Figures 3.5 to 3.9. Propositional networks consist of a set of nodes that represent sources of information, agents, and objects etc. that are linked through specific causal paths. From this network, it is possible to identify required information and possible options relevant to this incident. The concept behind using a propositional network in this manner is that it represents the 'ideal' collection of knowledge for the scenario. As the incident unfolds, so participants will have access to more of this knowledge (either through communication with other agents or through recognising changes in the incident status). Consequently, within this propositional network, Situation Awareness can be represented as the change in weighting of links. Propositional networks were developed for the overall scenario and also the incident phases identified during the CDM analysis. The propositional networks indicate which of the knowledge objects are active (that is, agents are using them) during each incident phase. The shaded

Table 3.11   CDM Phase 2: Deal with switching requests

Goal Specification	Obtain confirmation from NOC that planned isolation is still required.
Cue Identification	Approaching time for planned isolation.
	Switching phone rings throughout building.
	Airblast circuit breakers (accompanied by sirens) can be heard to operate remotely (more so in Barking 275 than Barking C 132).
Expectancy	Yes – routine planned work according to fixed procedures.
Conceptual Model	Wokingham have performed remote isolations already.
	Circuit configured ready for local isolation.
Uncertainty	Physical verification of apparatus always required (DBI – don't believe it).
Information	Proceduralised information from NOC – circuit, location, time, actions required etc.
	Switching log.
Situation Awareness	Switching log.
	Physical status of apparatus.
	Planning documentation.
	Visual or verbal information from substation personnel.
Situation Assessment	Planning documentation used only occasionally.
Options	Refusal of switching request.
	Additional conditions to switching request.
Stress	Some time pressure.
Choice	Yes – highly proceduralised anyway.
Analogy	Yes – routine activity.

nodes in the propositional networks represent unactivated knowledge objects (that is, knowledge is available but is not required nor is it being used). The darker nodes represent active (or currently being used) knowledge objects.

The results from the EAST analysis indicate that the key nodes in the Barking scenario network were the NOC operator and the SAP at Barking substation. The majority of the communications observed occurred between the NOC and the SAP, and the operational loading figures indicate that the SAP at Barking conducted the majority of the work required, on instruction from the NOC operator. Over half of all the tasks involved were identified as teamwork tasks, and a medium level of co-ordination was calculated between the agents involved. The EAST analysis indicated that the NOC operator assumes the role of commander during the scenario, and all work is undertaken only on his instruction.

Table 3.12 CDM Phase 3: Perform Isolation

Goal Specification	Ensure it is safe to perform local isolation.
	Confirm circuits/equipment to be operated.
Cue Identification	Telecontrol displays/circuit loadings.
	Equipment labels.
	Equipment displays.
	Other temporary notices.
Expectancy	Equipment configured according to planned circuit switching.
	Equipment will function correctly.
Conceptual Model	Layout/type/characteristics of circuit.
	Circuit loadings/balance.
	Function of equipment.
Uncertainty	Will equipment physically work as expected (will something jam etc.).
	Other work being carried out by other parties (e.g., EDF).
Information	Switching log.
	Visual and verbal information from those undertaking the work.
Situation Awareness	Physical information from apparatus and telecontrol displays.
Situation Assessment	All information used.
Options	Inform NOC that isolation cannot be performed/other aspects of switching instructions cannot be carried out.
Stress	Some time pressure.
	Possibly some difficulties in operating or physically handling the equipment.
Choice	Yes – proceduralised within equipment types. Occasional non-routine activities required to cope with unusual/unfamiliar equipment, or equipment not owned by NGT.
Analogy	Yes – often. Except in cases with unfamiliar equipment.

Related methods
The EAST methodology comprises a number of HF methods, including HTA, observation, CDA, CUD, OSDs, SNA, CDM and propositional networks. The authors have also developed the WESTT methodology, which can be used to automate a large portion of the EAST analysis procedure.

Approximate training and application times
The EAST methodology is conceptually simple, however due to the number of component methods involved, requires a lengthy training session. Some prior experience in the application of HF methods is required. For the recent applications

Table 3.13 CDM Phase 4: Report back to NOC

Goal Specification	Inform NOC of isolation status.
Cue Identification	Switching telephone.
	NOC operator answers.
Expectancy	NOC accepts.
Conceptual Model	Manner in which circuit is now isolated.
	Form of procedures.
Uncertainty	No – possibly further instructions, possibly mismatches local situation and remote displays in NOC.
Information	Switching log.
Situation Awareness	Verbal information from NOC.
	Switching log.
Situation Assessment	Yes – all information used.
Options	No (raise or add on further requests etc. to the same call?)
Stress	No.
Choice	Yes – highly proceduralised.
Analogy	Yes – frequently performed activity.

of EAST described in this chapter, a training session lasting approximately eight hours was given to the analysts involved. Due to the exhaustive nature of the EAST methodology, and the complex nature of C4i activity, the typical application time is high. The construction of the HTA and the OSD for C4i scenarios is a particularly lengthy process. Conducting an EAST analysis manually can take up to two weeks for large, complex scenarios. The application time is significantly reduced by the provision of the WESTT software package, which automates a large portion of the EAST analysis process.

Reliability and validity

Intra and inter reliability may be questionable in some instances. For example, different analysts may make different interpretations of the OSD symbols in the OSD glossary. The use of multiple methods within the EAST framework heightens the potential for unreliability. It is difficult to assess the validity of the data obtained during an EAST analysis. Typically, SMEs are used during data collection, and so a high level of validity is assumed.

Congruent validity is favourable due to effective methods integration. Construct validity is also good, as component methods relate well to the core descriptive constructs of C4i (Walker et al., 2006b). Construct and face validity are also good because many of the features of interest are observable and manifest in the

68 *Modelling Command and Control*

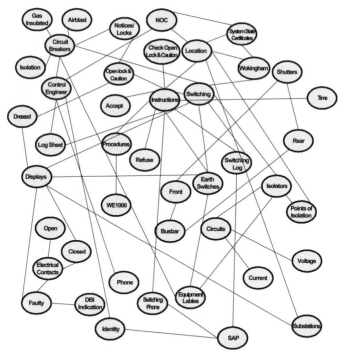

Figure 3.5 **Propositional network for objects referred to in CDM tables**

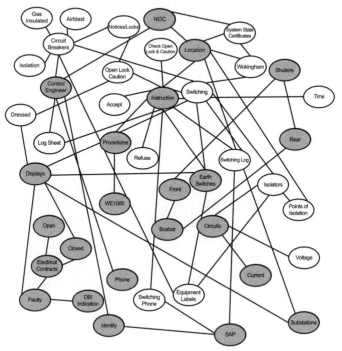

Figure 3.6 **Propositional network for CDM phase one**

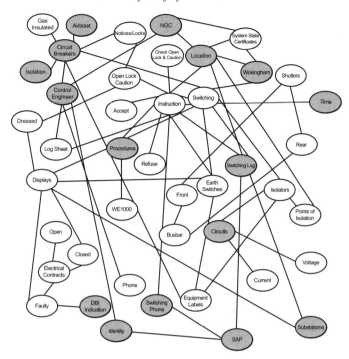

Figure 3.7 Propositional network for CDM phase two

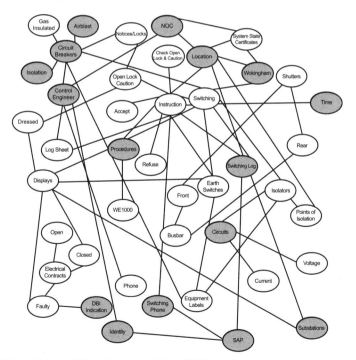

Figure 3.8 Propositional network for CDM phase three

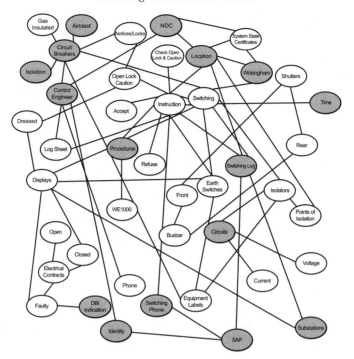

Figure 3.9 Propositional network for CDM phase four

environment of interest. The CDM complements this approach by tapping into the 'implicit' parts of the decision making process.

Tools needed
Due to the various component analyses involved in an EAST analysis, a number of different tools and software packages are required. For the observation and CDM phases of an EAST analysis, video and audio recording equipment is required. A word processing software package such as Microsoft Word is required to transcribe the observation and CDM data. The CDA is typically conducted using Microsoft Excel. A large part (for example, OSD and SNA) of the EAST procedure can be automated by using the WESTT software package. Alternatively, the OSD CUD, and propositional networks are typically drawn using Microsoft Visio. The SNA is conducted using the AGNA SNA software package.

Summary flowchart

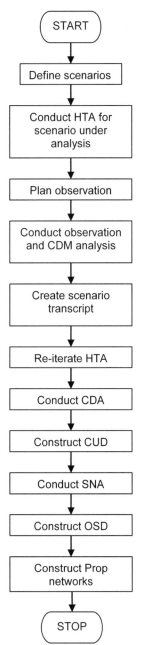

Observation

Background and applications
Observational techniques are used to gather data regarding the physical and verbal aspects of a particular task or scenario. Observational techniques are used to collect data regarding various aspects of system and task performance, such as data regarding the tasks catered for by the system, the individuals performing the tasks, the tasks themselves (task steps and sequence), errors made, communications between individuals, the technology used by the system in conducting the tasks (controls, displays, communication technology etc.), the system environment and the organisational environment. Observation has been extensively used in the HF community, and typically forms the beginning of an analysis effort.

Domain of application
Generic.

Procedure and advice

Step 1: Define the objective of the analysis The first step in observational analysis involves defining the aims and objectives of the observation. This should include determining which product or system is under analysis, in which environment the observation will take place, which user groups will be observed, what type of scenarios will be observed, and what data are required. Each point should be clearly defined and stated before the process continues.

Step 2: Define the scenario(s) Once the aims and objectives of the analysis are clearly defined, the scenario(s) to be observed should be defined and described further. For example, when conducting an observational analysis of control room operation, which type of scenario is required should be clearly defined. Is normal operation under scrutiny or is the analysis focussed upon operator interaction and performance under emergency situations. The exact nature of the required scenario(s) should be clearly defined by the observation team. It is recommended that a HTA is conducted for the task or scenario under analysis.

Step 3: Observation plan   Once the aim of the analysis is defined and also the type of scenario to be observed is determined, the analysis team should proceed to plan the observation. The team should consider what they are hoping to observe, what they are observing, and how they are going to observe it. Depending upon the nature of the observation, access to the system in question should be gained first. This may involve holding meetings with the establishment in question, and is typically a lengthy process. Any recording tools should be defined and also the length of observations should be determined. Placement of video and audio recording equipment should also be considered. To make things easier, a walkthrough of the system/environment/scenario under analysis is recommended. This allows the analyst(s) to become familiar with the task in terms of time taken, location and also the system under analysis.

Step 4: Pilot observation   In any observational study a pilot or practice observation is crucial. This allows the analysis team to assess any problems with the data collection, such as noise interference or problems with the recording equipment. The quality of data collected can also be tested and also any effects of the observation upon task performance can be assessed. If major problems are encountered, the observation may have to be re-designed. Steps 1 to 4 should be repeated until the analysis team are happy that the quality of the data collected will be sufficient for their study requirements.

Step 5: Conduct Observation   Once the observation has been designed, the team should proceed with the observation(s). Observation length and timing are dependent upon the scope and requirements of the analysis. Once the required data are collected, the observation should stop and step 6 should be undertaken.

Step 6: Create observation transcript   Once the observation is complete, the analyst(s) should construct an observation transcript. An extract of an observation transcript from an observation of an energy distribution switching scenario (Salmon et al., 2004b; 2004c) is presented in Table 3.14. The observational transcript should contain a description of all activity observed during the analysis, including the agents involved, the time, the technology used and also any additional notes (for example, purpose of activity).

Step 7: Data analysis   Once the observation is complete, and the observation transcripts have been constructed, the data analysis procedure can begin. Depending upon the analysis requirements, the team should analyse the data in the format that is required, such as frequency of tasks, verbal interactions, sequence of tasks etc. When analysing visual data, typically user behaviours are coded into specific groups. The software package Observer™ is used to aid the analyst in this process.

Step 8: Further analysis   Once the initial process of transcribing and coding the observational data is complete, further analysis of the data begins. Depending upon the nature of the analysis, observation data are used to inform a number of different HF analyses, such as task analysis, error analysis and communications analysis.

Table 3.14 Observation transcript extract (Source: Salmon et al., 2004b)

Time	Process	Comms	Location	Notes
09:54	GA & DW engage in general pre-amble about the forthcoming switching ops. Considering asking the Operations Centre in Wokingham to switch out SGT1A1B early so that SGT5 can be done at the same time?	Person to person	Barking 275Kv switch-house	
10:15	Wokingham call Barking ask if they still want isolation – GA confirms yes.	Telephone		
10:19	Switching phone rings throughout building. NOC? Wants 132Kv busbar opened [GA].	Green Telephone		
10:40	SGT1A1B waiting to be handed over to NOC. Delay.			
10:40	Wokingham report to Barking 275 [GA] complication with EDF. EDF want to reselect circuits at the last minute at another substation due to planned shutdown of SGT1A.	Telephone		Wokingham contact EDF to confirm switching (as they did with Barking 275 at 10:15). EDF report a problem. Wokingham pass this onto Barking 275.
10:53	GA and DW discuss EDF problem. Can DW reconfigure circuits in Barking West (33Kv)?	Person to person		
10:53	GA contact Wokingham. Confusion as to what circuits need reconfiguring and who can do it. GA talks to DW at the same time. Decided that DW can reconfigure circuits. Wokingham give GA name and phone number of EDF contact.	GA/DW Person to Person, Wokingham Telephone		This is an unplanned measure – now need to go to Barking West 33Kv to reconfigure local electricity supply circuits.
10:58	GA and DW discuss plans for Barking West 33. Discuss who owns what, safety rules etc. Also discuss and decide order of subsequent site visits (might have to go to Barking West 33 twice).			
11:04	GA & DW Waiting to travel to Barking West.			

Typically, observational data are used to develop a task analysis (for example, HTA) of the task or scenario under analysis.

Step 9: Participant feedback   Once the data have been analysed and conclusions have been drawn, the participants involved should be provided with feedback of some sort. This could be in the form of a feedback session or a letter to each participant. The type of feedback used is determined by the analysis team.

Advantages

- Observation technique data provides a 'real life' insight into man-machine, and team interaction.
- Various data can be elicited from an observational study, including task sequences, task analysis, error data, task times, verbal interaction and task performance.
- Observation has been used extensively in a wide range of domains.
- Observation provides objective information.
- Detailed physical task performance data are recorded, including social interactions and any environmental task influences (Kirwan and Ainsworth, 1992).
- Observation is excellent for the initial stages of the task analysis procedure.
- Observation analysis can be used to highlight problems with existing operational systems. It can be used in this way to inform the design of new systems or devices.
- Specific scenarios are observed in their 'real world' setting.
- Observational data are used as the core input into numerous HF analysis techniques, such as human error identification techniques (SHERPA), task analysis (HTA), communications analysis (Comms Usage Diagrams), and charting techniques (operator sequence diagrams).

Disadvantages

- Observational techniques are intrusive to task performance. Knowing that they are being watched tends to elicit new and different behaviours in participants. For example, when observing control room operators, they may perform exactly as their procedures say they should. However, when not being observed, the same control room operators may perform completely differently, using short-cuts and behaviours that are not stated in their procedures. This may be due to the fact that the operators do not wish to be caught bending the rules in any way, for example bypassing a certain procedure. This effect on behaviour can bias the observational data that is obtained.
- Observational techniques are time consuming in their application, particularly the data analysis procedure. Kirwan and Ainsworth (1992) suggest that when conducting the transcription process, one hour of recorded audio data takes one analyst approximately eight hours to transcribe.

- Cognitive aspects of the task under analysis are not elicited using observational techniques. Verbal protocol analysis is more suited for collecting data on the cognitive aspects of task performance.
- An observational study can be both difficult and expensive to set up and conduct. Gaining access to the required establishment is often extremely difficult and very time consuming. Observational techniques are also costly, as they require the use of expensive recording equipment (digital video camera, audio recording devices).
- Causality is a problem. Errors can be observed and recorded during an observation but why the errors occur may not always be clear.
- The analyst has a very low level of experimental control.
- In most cases, a team of analysts is required to perform an observation study. It is often difficult to acquire a suitable team with sufficient experience in conducting observational studies.

Example

An observational analysis of a fire service training scenario was conducted as part of an analysis of existing C4i activity in civil domains (Baber et al., 2004). Three observers observed the fire training service-training scenario, 'Hazardous chemical spillage at remote farmhouse'. The three observers sat in on the exercise and recorded the discussion of the participants. The notes from the discussion were then collated into a combined timeline of the incident. This timeline, and the notes taken during the exercise, then formed the basis for the following HF analyses.

- hierarchical task analysis
- operator sequence diagram
- social network analysis
- co-ordination demand analysis
- comms usage diagram
- critical decision method.

The exercise involved a combination of focus group discussion with paired activity in order to define appropriate courses of action to deal with the specified incident. The class facilitator provided the initial description of an incident, that is, a report of possible hazardous materials on a remote farm, and then added additional information as the incident unfolded, for example, reports of casualties, problems with labelling on hazardous materials etc. The exercise was designed to encourage experienced fire-fighters to consider risks arising from hazardous materials and the appropriate courses of action they would need to take, for example in terms of protective equipment, incident management, information seeking activities etc. From the data obtained during the observation, an event flowchart was constructed, which acted as the primary input to the analysis techniques used. The event flowchart is presented in Figure 3.10.

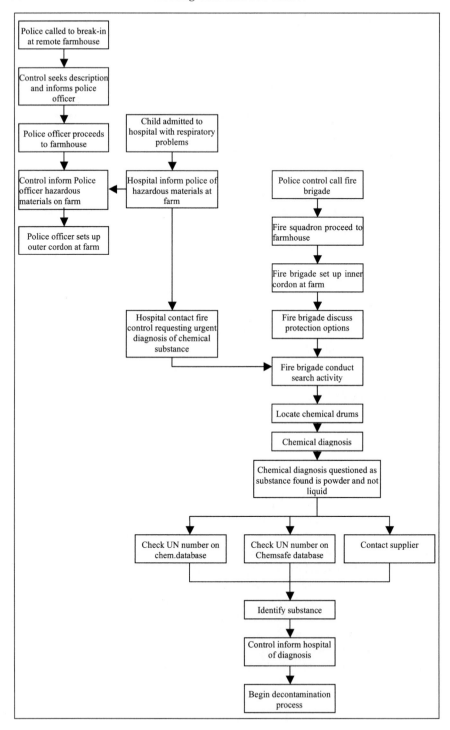

Figure 3.10 Hazardous chemical spillage event flowchart

Related methods
There are a number of different observational techniques, including indirect observation, participant observation and remote observation. Other related techniques include verbal protocol analysis, critical decision method, applied cognitive task analysis, walkthroughs and cognitive walkthroughs. All of these techniques require some sort of task observation. Observational data are also used as an input to numerous HF techniques, such as task analysis, human error identification, and charting techniques.

Approximate training and application times
Whilst the training time for an observational analysis is low (Stanton and Young, 1999), the application time is normally high. The data analysis stage can be particularly time consuming, as discussed above.

Reliability and validity
Observational analysis is beset by a number of problems that can potentially affect the reliability and validity of the technique. According to Baber and Stanton (1996) problems with causality, bias (in a number of forms), construct validity, external validity and internal validity can all arise unless the correct precautions are taken. Whilst observational techniques possess a high level of face validity (Drury, 1990) and ecological validity (Baber and Stanton, 1996), analyst or participant bias can adversely affect the reliability and validity of the techniques.

Tools needed
For a thorough observational analysis, the appropriate visual and audio recording equipment is necessary. Observational studies can be conducted using pen and paper, however this is not recommended, as crucial parts of data are often not recorded. For the data analysis process, a PC with the Observer™ software is required.

Summary flowchart

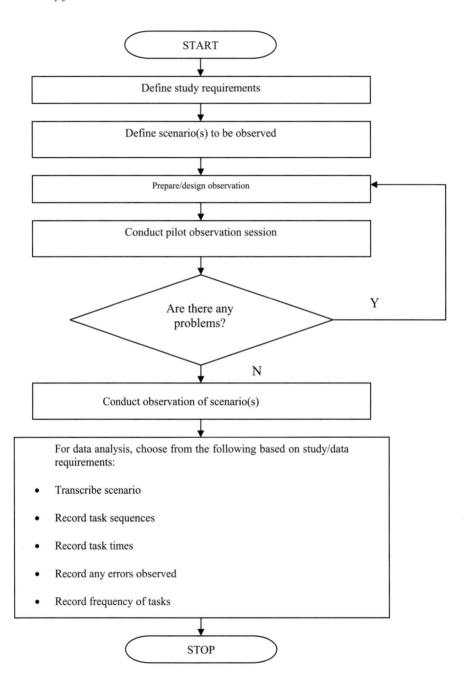

HTA - Hierarchical task analysis

Background and applications

Hierarchical task analysis (HTA) (Annett, 2004) is the most popular task analysis technique and has become perhaps the most widely used of all HF techniques. HTA involves describing the task under analysis through the break down of the task into a hierarchy of goals, sub goals, operations and plans. Goals are the unobservable task goals associated with the task in question, operations are the observable behaviours required to accomplish the task goals, and plans are the unobservable decisions and planning made on behalf of the operator. The end result is an exhaustive description of task activity. HTA has been applied across a wide spectrum of domains, including the process control and power generation industries (Annett, 2004), emergency services (Baber et al., 2004) military applications (Kirwan and Ainsworth, 1992; Ainsworth and Marshall, 1998/2000), civil aviation (Marshall et al., 2003), driving, public technology (Stanton and Stevenage, 1998) and even retail (Shepherd, 2001). One of the main reasons behind the enduring popularity of HTA is its flexibility and the scope for further analysis of the sub-goal hierarchy that it offers to the HF practitioner. The majority of HF analysis methods either require an initial HTA of the task(s) under analysis as their input, or at least are made significantly easier through the provision of a HTA. HTA acts as an input into numerous HF analyses techniques, such as human error identification (HEI), allocation of function, workload assessment, interface design and evaluation, and many more. In a review of ergonomics texts, Stanton (2006) highlights at least 12 additional applications to which HTA has been put, including interface design and evaluation, training, allocation of functions, job description, work organisation, manual design, job aid design, error prediction and analysis, team task analysis, workload assessment and procedure design.

Domain of application

HTA was originally developed for the chemical processing and power generation industries (Annett, 2004). However the technique is generic and can be applied in any domain.

Procedure and advice

Step 1: Define task under analysis   The first step in conducting a HTA is to clearly define the task(s) under analysis. As well as identifying the task under analysis, the purpose of the task analysis effort should also be defined. For example, Marshall et al. (2003) conducted a HTA of a civil aircraft landing task in order to predict design induced error for the flight task in question.

Step 2: Data collection process   Once the task under analysis is clearly defined, specific data regarding the task should be collected. The data collected during this process is used to inform the development of the HTA. Data regarding the task steps involved, the technology used, interaction between man and machine and team members, decision making and task constraints should be collected. There are a number of ways to collect these data, including observations, interviews, and questionnaires.

The technique used is dependent upon the analysis effort and the various constraints imposed, such as time and access constraints. Once sufficient data regarding the task under analysis are collected, the development of the HTA should begin.

Step 3: Determine the overall goal of the task The overall task goal of the task under analysis should first be specified at the top of the hierarchy, that is 'Land aircraft X at New Orleans Airport using the Auto-land system' (Marshall et al., 2003), 'Boil kettle', or 'Listen to in-car entertainment' (Stanton and Young, 1999).

Step 4: Determine task sub-goals Once the overall task goal has been specified, the next step is to break this overall goal down into meaningful sub-goals (usually four or five but this is not rigid), which together form the tasks required to achieve the overall goal. In the task, 'Land aircraft X at New Orleans Airport using the Auto-land system' (Marshall et al., 2003), the overall goal of landing the aircraft was broken down into the sub-goals, 'Set up for approach', 'Line up aircraft for runway' and 'Prepare aircraft for landing'. In a HTA of a Ford in-car radio (Stanton and Young, 1999) the overall task goal, 'Listen to in-car entertainment', was broken down into the following sub-goals, 'Check unit status', 'Press on/off button', 'Listen to the radio', 'Listen to cassette', and 'Adjust audio preferences'.

Step 5: Sub-goal decomposition Next, the analyst should break down the sub-goals identified in step four into further sub-goals and operations, according to the task step in question. This process should go on until an appropriate operation is reached. The bottom level of any branch in a HTA will always be an operation. Whilst everything above an operation specifies goals, operations actually say what needs to be done. Thus operations are actions to be made by the operator in order to achieve the associated goal. For example, in the HTA of the flight task 'Land aircraft X at New Orleans Airport using the Auto-land system' (Marshall et al., 2003), the sub-goal 'Reduce airspeed to 210 Knots' is broken down into the following operations, 'Check current airspeed' and 'Dial the Speed/MACH selector knob to enter 210 on the IAS/MACH display'.

Step 6: Plans analysis Once all of the sub-goals and operations have been fully described, the plans need to be added. Plans dictate how the goals are achieved. A simple plan would say Do 1, then 2, and then 3. Once the plan is completed, the operator returns to the super-ordinate level. Plans do not have to be linear and can come in any form such as Do 1, or 2 and 3. The different types of plans used are presented in Table 3.15.

Advantages

- HTA is a technique that is both easy to learn and to implement, but can be time consuming to apply.
- The output of a HTA is extremely useful and forms the input for numerous HF analyses, such as error analysis, interface design and evaluation and allocation of function analysis.
- Quick to use in most instances.

- Comprehensive technique covers all sub-tasks of the task in question.
- HTA has been used extensively in a wide range of contexts.
- Conducting a HTA gives the user a great insight into the task under analysis.
- HTA is an excellent technique to use when requiring a task description for further analysis. If performed correctly, the HTA should depict everything that needs to be done in order to complete the task in question.
- The technique is generic and can be applied to any task in any domain.
- Tasks can be analysed to any required level of detail, depending on the purpose.

Disadvantages

- Provides mainly descriptive information rather than analytical information.
- HTA contains little that can be used directly to provide design solutions.
- HTA does not cater for the cognitive components of the task under analysis.
- The technique may become laborious and time consuming to conduct for more complex and larger tasks.
- The initial data collection phase is time consuming and requires the analyst to be competent in a variety of HF techniques, such as interviews, observations and questionnaires.
- The reliability of the technique may be questionable in some instances. For example, for the same task, different analysts may produce very different sub-goal hierarchies.
- HTA has often been labelled as an art rather than a science.

Example

The following HTA was conducted as part of an analysis of C4i activity in the naval warfare domain (Stewart et al., 2004). The HTA describes the activity of the principal warfare officer (PWO) during a subsurface attack scenario. An observation of a naval training exercise and an interview with a SME were the primary input into the development of the HTA. The HTA is presented below.

Table 3.15   Example HTA plans

Plan	Example
Linear	Do 1 then 2 then 3
Non-linear	Do 1, 2 and 3 in any order
Simultaneous	Do 1, then 2 and 3 at the same time
Branching	Do 1, if X present then do 2 then 3, if X is not present then EXIT
Cyclical	Do 1 then 2 then 3 and repeat until X
Selection	Do 1 then 2 or 3

Glossary

PWO = Principal Warfare Officer
ASWD = Active Systems Weapons Director
PSD = Passive System Director
PC = Picture Compilers
HC = Helicopter
ASonar = Active Sonar team
PSonar = Passive Sonar team
CO = Commanding Officer

0. Maintain control of sea/air environment

Plan 0. Do 1 then 2 Is object Hostile No then EXIT, Yes then 3 then 4 then 5 then 6 (repeat until target is destroyed)
Has target responded? Yes then 7 No then EXIT

1. Plan resources and strategy

Plan 1. Do 1.1 then 1.2 then 1.3 then 1.4 then 1.5 then 1.6 then 1.7 then 1.8
    1.1. (AWO + ASWD + PSD) Determine mission plans
    Plan 1.1.1. Do 1.1.1 - 1.1.5 as appropriate then 1.1.6. EXIT
        1.1.1. determine overt plan
        1.1.2. determine covert plan
        1.1.3. determine day plan
        1.1.4. determine night plan
        1.1.5. determine satellite passes
        1.1.6. Choose plan
    1.2. (AWO + ASWD + PSD) Determine immediate priorities
    1.3. (AWO) Conduct general briefing and discussion
    1.4. (PWO + ASWD + PSD) Conduct threat analysis
    1.5. (PWO + ASWD + PSD) Balance threats to assets over given timeline
    1.6. (PWO + ASWD + PSD + HC) Draw up plan
    1.7. (PWO + ASWD + PSD + HC) Order ind specialists to write up plan execution instructions
    1.8. (PWO + ASWD + PSD + HC) Compile operational order to guide activity

2. Detect, identify, evaluate and allocate area movement

Plan 2. Do 2.1 Are object(s) detected, No then EXIT, Yes then 2.2 then EXIT

    2.1. (Sonar teams + ASD + PSD) Detect targets
    2.2. (Sonar teams + ASD + PSD) Identify/Classify underwater objects
    Plan 2.2. Do 2.2.1 - 2.2.7 Based upon object type
        2.2.1. (Sonar teams + ASD + PSD) Classify as friendly
        2.2.2. (Sonar teams + ASD + PSD) Classify as neutral
        2.2.3. (Sonar teams + ASD + PSD) Classify as unknown
        2.2.4. (Sonar teams + ASD + PSD) Classify as unknown assumed friendly
        2.2.5. (Sonar teams + ASD + PSD) Classify as unknown assumed enemy

2.2.6. (Sonar teams + ASD + PSD) Classify as hostile
2.2.7. (Sonar teams + ASD + PSD) Classify as hostile

3. Respond to targets

Plan 3. Do 3.1 then 3.2 then 3.3 then 3.4 then 3.5 then 3.6 and/or 3.7 if engaging target then 3.8
then EXIT, if not engaging target then EXIT
Can target be evaded? Yes then 3.5 then EXIT,
No then 3.6 then EXIT
    3.1. (PWO + ASonar) Determine target geography
    3.2. (PWO) Determine target capability
    3.3. (PWO + CO) Determine response (attack and/or evade)
    3.4. Conduct threat identification
    Plan 3.4. Do 3.4.1 then 3.4.2 then 3.4.3 then 3.4.4 then 3.4.5 EXIT
        3.4.1. (PWO) Assess weapons capability
        3.4.2. (PWO) Assess posture intent
        3.4.3. (PWO) Assess radar status
        3.4.4. (PWO) Assess target range
        3.4.5. (PWO) Assess risks from target
    3.5. (PWO) Determine defence type (decoys or missiles)
    3.6. (PWO) Evade target
    3.7. (PWO) Engage targets
Plan 3.7. Do 3.7.1 then 3.7.2 then 3.7.3 then 3.7.4 then 3.7.5 then 3.7.6 then EXIT
        3.7.1. (PWO) Match target to weapon system
        3.7.2. (PWO) Ensure that all targets are engaged
        3.7.3. (PWO) Check that no target is over engaged
        3.7.4. (PWO) Check that no target is unengaged
        3.7.5. (PWO) Order weapons systems to targets
        3.7.6. (PWO) Open comms to battle group on target track allocation
    3.8. Conduct target allocation
    Plan 3.8. Do 3.8.1 then 3.8.2, then 3.8.3 then 3.8.4 EXIT
        3.8.1. (PWO) Prioritise weapons
        3.8.2. (PWO) Allocate appropriate weapons to target
        3.8.3. (PWO) Identify new target data
        3.8.4. (PWO) Reassess allocation of weapons

4. Control external resources

Plan 4. Do 4.1 and/or 4.2 and/or 4.3 then 4.4 then EXIT
    4.1. (PWO) Get intelligence report
    4.2. (PWO) Get report from airborne assets
    4.3. (PWO) Get report from strategic assets
    4.4. (PWO) Direct resources to deal with threat
    Plan 4.4. Do 4.4.1 then 4.4.2 then 4.4.3 then 4.4.4 then 4.4.5 then 4.4.6 then 4.4.7 then EXIT
        4.4.1. (PWO) Identify threat sector
        4.4.2. (PWO) Posture platform to meet threat
        4.4.3. (PWO) Direct fighters to threat sector
        4.4.4. (PWO) Put stations in rings around high value units

4.4.5. (PWO) Check allocation of fighter stations as threat develops
4.4.6. (PWO) Reallocate fighters stations if required
4.4.7. (PWO) Allocate target engagement to E3D if necessary

5. Posture platform in response to attack

Plan 5. Do 5.1 then 5.2 then 5.3 then 5.4 then 5.5 and/or 5.6 as appropriate then 5.7 then 5.8 then 5.9 then EXIT
    5.1. (PWO) Determine likely submarine weapon
    5.2. (PWO) Determine point of incoming threat
    5.3. (PWO) Countdown range from target
    5.4. (PWO) Identify missile mode (e.g. radar bandwidth used etc)
    5.5. (PWO) Change platform heading
    Plan 5.5. Do 5.5.1 then 5.5.2 then 5.5.3 then EXIT
        5.5.1. (PWO) Check current platform heading
        5.5.2. (PWO) Change platform heading to give best target acquisition
        5.5.3. (PWO) Change platform heading to give best decoy deployment
    5.6. (PWO) Change current speed
    Plan 5.6. Do 5.6.1 then 5.6.2 then EXIT
        5.6.1. (PWO) Check current speed
        5.6.2. (PWO) Change speed as appropriate
    5.7. (PWO) Select appropriate sonar mode
    5.8. (PWO) Adjust sonar mode
    5.9. (PWO) Build picture using alternative resources
    Plan 5.9. Do 5.9.1 and/or 5.9.2 and/or 5.9.3 then 5.9.4
        5.9.1. (PWO) Use reconnaissance data
        5.9.2. (PWO) Use alternative floating asset data
        5.9.3. (PWO) Use strategic intelligence
        5.9.4. (PWO) Determine picture

6. (PWO + ASonar, PSonar + HC + other ships) Conduct damage assessment

Plan 6. Do 6.1 then 6.2 then 6.3 then 6.4 then EXIT
    6.1. Listen to torpedo running
    6.2. Check sonar
    6.3. Determine damage

7. (PWO + ASWD) Defend ship

Plan 7. Do 7.1 then 7.2 then 7.3 then 7.4 then 7.5 then EXIT
    7.1. Determine threat type
    7.2. Check decoy look up tables
    7.3. Select appropriate decoy type
    7.4. Instruct decoy management to launch decoys
    7.5. Deploy decoys

Related methods

HTA is widely used in HF and often forms the first step in a number of analyses, such as HEI, HRA and mental workload assessment. In a review of ergonomics texts, Stanton (2006) highlights at least 12 additional applications to which HTA has been put, including interface design and evaluation, training, allocation of functions, job description, work organisation, manual design, job aid design, error prediction and analysis, team task analysis, workload assessment and procedure design. As a result HTA is perhaps the most commonly used HF method, and is typically used as the starting point or basis of any HF analysis.

Approximate training and application times
The training time for the HTA method is largely dependent upon on the skill of the analysts involved. It is often said that using HTA is more of an art than a science, and the training time required reflects this. Some analysts may pick up the method immediately, whilst others may use the technique repeatedly without becoming skilled in its use. It is therefore estimated that the training time for HTA is high. A survey by Ainsworth and Marshall (1998/2000) found that the more experienced practitioners produced more complete and acceptable analyses. Stanton and Young (1999) report that the training and application time for HTA is easy to learn, but can be time consuming to apply. The application time associated with HTA is dependent upon the size and complexity of the task under analysis. For large, complex tasks, the application time for HTA would be high.

Reliability and validity
According to Annett (2004), the reliability and validity of HTA is not easily assessed. From a comparison of 12 HF techniques, Stanton and Young (1999) reported that, the technique achieved an acceptable level of validity but a poor level of reliability. The reliability of the technique is certainly questionable. It seems that different analysts, with different experience may produce entirely different analyses for the same task (intra-analyst reliability). Similarly, the same analyst may produce different analyses on different occasions for the same task (inter-analyst reliability).

Tools needed
HTA can be carried out using only a pencil and paper. The HTA output can be developed and presented in a number of software applications, such as Microsoft Visio, Microsoft Word and Microsoft Excel. A number of HTA software tools also exist, such as the HFI DTC's HTA tool.

Summary flowchart

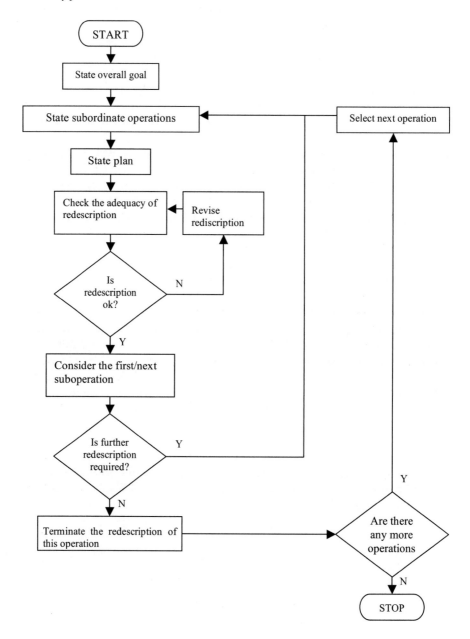

CDA – Co-ordination demands analysis

Background and application

Co-ordination demands analysis (CDA) is used to rate the co-ordination between agents involved in teamwork activity. CDA is used to identify the extent to which team members have to work with each other in order to accomplish the task(s) under analysis. The CDA procedure allows analysts to identify the operational skills required within team tasks, but also the teamwork skills needed for smooth coordination among team members. In its present usage teamwork skills are extracted from a HTA (Annett, 2004) and rated on a scale of 1 (low) to 3 (high) against each behaviour in the co-ordination behaviour taxonomy presented in Table 3.16 (Source: Burke, 2005). From these individual ratings a 'total coordination' figure for each teamwork task step is derived.

Table 3.16   CDA teamwork taxonomy

Coordination Dimension	Definition
Communication	Includes sending, receiving, and acknowledging information among crewmembers.
Situational Awareness (SA)	Refers to identifying the source and nature of problems, maintaining an accurate perception of the aircraft's location relative to the external environment, and detecting situations that require action.
Decision Making (DM)	Includes identifying possible solutions to problems, evaluating the consequences of each alternative, selecting the best alternative, and gathering information needed prior to arriving at a decision.
Mission analysis (MA)	Includes monitoring, allocating, and coordinating the resources of the crew and aircraft; prioritising tasks; setting goals and developing plans to accomplish the goals; creating contingency plans.
Leadership	Refers to directing activities of others, monitoring and assessing the performance of crew members, motivating members, and communicating mission requirements.
Adaptability	Refers to the ability to alter one's course of action as necessary, maintain constructive behaviour under pressure, and adapt to internal or external changes.
Assertiveness	Refers to the willingness to make decisions, demonstrating initiative, and maintaining one's position until convinced otherwise by facts.
Total Coordination	Refers to the overall need for interaction and coordination among crew members.

Source: Burke, 2004

Domain of application

The CDA technique is generic and can be applied to any task that involves teamwork or collaboration.

Procedure and advice

Step 1: Define task(s) under analysis   The first step in a CDA is to define the task or scenario that will be analysed. This is dependent upon the focus of the analysis. It is recommended that if team coordination in a particular type of system (for example command and control) is under investigation, then a set of scenarios representative of team performance in the system should be used. If time and financial constraints do not allow this, then a task that is as representative as possible of team performance in the system under analysis should be used.

Step 2: Select appropriate teamwork taxonomy   Once the task(s) under analysis are defined, an appropriate teamwork taxonomy should be selected. Again, this may depend upon the purpose of the analysis. However, it is recommended that the taxonomy used covers all aspects of teamwork in the task under analysis.

Step 3: Data collection phase   Typically, data regarding the task(s) under analysis is collected using observation and/or interviews. Specific data regarding the task under analysis should be collected during this process, including information regarding each task step, each team member's roles, and communications made. It is recommended that video and audio recording equipment are used to record any observations or interviews conducted during this process.

Step 4: Conduct a HTA for the task under analysis   Once sufficient data regarding the task under analysis has been collected, a HTA should be conducted. The HTA should represent a breakdown of the task under analysis in terms of goals, sub-goals, operations and plans.

Step 5: Construct CDA rating sheet   Once a HTA for the task under analysis is completed, a CDA rating sheet should be created. The rating sheet should include a column containing each bottom level task step as identified by the HTA. The teamwork behaviours from the taxonomy should run across the top of the table. An extract of a CDA rating sheet is presented in Table 3.17.

Step 6: Taskwork/Teamwork classification   Once the CDA rating sheet is complete, the analyst(s) should then determine which of the bottom level task steps from the HTA involve taskwork and which involve teamwork. Only those task steps defined as teamwork task steps should be rated according to the teamwork taxonomy. Those task steps that are conducted individually involving no collaboration are classified as taskwork, whilst those task steps that are conducted collaboratively, involving more than one agent are classified as teamwork.

Table 3.17 CUD summary table

Agent	Telephone comms	Mobile phone comms	PC comms	Total
NOC operator	11			11
SAP/AP at Tottenham	8	1		9
SAP/AP at Waltham Cross	5			5
SAP/AP at Brimsdown	7			7
Overhead line party contact		1		1
WOK operator	1			1
Total	**32**	**2**		**34**

Step 7: SME rating phase Appropriate SMEs should then rate the extent to which each teamwork behaviour is required during the completion of each teamwork task step. This involves presenting the task step in question and discussing the role of each taxonomy behaviour in the completion of the task step. An appropriate rating scale should be used, for example low (1), medium (2) and high (3).

Step 8: Calculate summary statistics Once all of the teamwork task steps have been rated according to the teamwork taxonomy, the final step is to calculate appropriate summary statistics. In its present usage, a total co-ordination value and mean co-ordination value for each teamwork task step are calculated. The mean co-ordination is simply an average of the ratings for the teamwork behaviours for the task step in question. A mean overall co-ordination value for the entire scenario is also calculated.

Advantages

- Offers a rating of co-ordination between team members for each teamwork based task step.
- The output of a CDA is very useful, offering an insight into the use of teamwork behaviours and also a rating of team coordination and its components.
- CDA is particularly useful for the analysis of C4i activity.
- The teamwork taxonomy presented by Burke (2005) covers all aspects of team performance and coordination. The taxonomy is also generic, allowing the technique to be used in any domain without modification.
- CDA provides a breakdown of team performance in terms of task steps and the level of co-ordination required.

- The technique is generic and can be applied to teamwork scenarios in any domain.

Disadvantages

- The CDA rating procedure is time consuming and laborious. The initial data collection phase and the creation of a HTA for the task under analysis also add further time to the analysis.
- For the technique to be used properly, the appropriate SMEs are required. It may be difficult to gain access to SMEs for the required period of time.
- Intra analyst and inter analyst reliability is questionable. Different SMEs may offer different teamwork ratings for the same task (intra analyst reliability), whilst SMEs may provide different ratings on different occasions.

Example

The following example is an extract of a CDA analysis of an energy distribution task (Salmon et al., 2004b; 2004c). The task involved the switching out of three circuits (SGT5 and SGT1A and 1B) at Barking 275Kv, 132Kv and 33Kv Substations. Circuit SGT5 was being switched out for the installation of a new transformer for the nearby channel tunnel rail link and SGT1A and 1B were being switched out for substation maintenance. Observational data from Barking substation and the NOC control room was used to conduct a HTA of the switching scenario. Each bottom level task in the HTA was then defined by the analyst(s) as either taskwork or teamwork. Each teamwork task was then rated using the CDA taxonomy on a scale of 1 (low) to 3 (high). An extract of the HTA for the task is presented below. An extract of the CDA has already been presented (see Table 3.5). The overall CDA results for this scenario have also been presented previously in Table 3.6.

Extract of NGC switching operations HTA

0. Co-ordinate and carry out switching operations on circuits SGT5. SGT1A and 1B at Bark s/s (Plan 0. Do 1 then 2 then 3, EXIT)

1. Prepare for switching operations (Plan 1. Do 1.1, then 1.2, then 1.3, then 1.4, then 1.5, then 1.6, then 1.7, then 1.8, then 1.9,then 1.10 EXIT)
    1.1. Agree SSC (Plan 1.1. Do 1.1.1, then 1.1.2, then 1.1.3, then 1.1.4, then 1.1.5, EXIT)
        1.1.1. (WOK) Use phone to Contact NOC
        1.1.2. (WOK + NOC) Exchange identities
        1.1.3. (WOK + NOC) Agree SSC documentation
        1.1.4. (WOK+NOC) Agree SSC and time (Plan 1.1.4. Do 1.1.4.1, then 1.1.4.2, EXIT)
            1.1.4.1. (NOC) Agree SSC with WOK
            1.1.4.2. (NOC) Agree time with WOK
        1.1.5. (NOC) Record and enter details (Plan 1.1.5. Do 1.1.5.1, then 1.1.5.2, EXIT)
            1.1.5.1. Record details on log sheet

1.1.5.2. Enter details into worksafe
1.2. (NOC) Request remote isolation (Plan 1.2. Do 1.2.1, then 1.2.2, then 1.2.3, then 1.2.4, EXIT)
    1.2.1. (NOC) Ask WOK for isolators to be opened remotely
    1.2.2. (WOK) Perform remote isolation
    1.2.3. (NOC) Check Barking s/s screen
    1.2.4. (WOK + NOC) End communications
1.3. Gather information on outage at transformer 5 at Bark s/s
(Plan 1.3. Do 1.3.1, then 1.3.2, then 1.3.3, then 1.3.4, EXIT)
    1.3.1. (NOC) Use phone to contact SAP at Bark
    1.3.2. (NOC + SAP) Exchange identities

Related methods

In conducting a CDA analysis, a number of other HF techniques are used. Data regarding the task under analysis is typically collected using observations and interviews. A HTA for the task under analysis is normally conducted, the output of which feeds into the CDA rating sheet. A Likert style rating scale is also normally used during the team behaviour rating procedure. Burke (2005) also suggests that a CDA should be conducted as part of an overall team task analysis procedure. The CDA technique has also recently been integrated with a number of other techniques (HTA, observation, comms usage diagram, social network analysis, operator sequence diagrams and propositional networks) to form the event analysis of systemic teamwork (EAST) methodology, which has been used to analyse C4i activity in a number of domains.

Approximate training and application times

The training time for the CDA technique is minimal, requiring only that the SMEs used understand each of the behaviours specified in the teamwork taxonomy and also the rating procedure. The application time is medium to high, depending upon the size and complexity of the scenario under analysis. In the CDA provided in the analysis, the ratings procedure alone took approximately four hours. This represents a medium application time.

Reliability and validity

There are no data regarding the reliability and validity of the technique available in the literature. Certainly both the intra analyst and inter analyst reliability of the technique may be questionable, and this may be dependent upon the type of rating scale used. For example it is estimated that the reliability may be low when using a scale of one to ten, whilst it may be improved using a scale of one to ten (low to high).

Tools needed

During the data collection phase, video (for example camcorder) and audio (for example recordable mini-disc player) recording equipment are required in order to make a recording of the task or scenario under analysis. Once the data collection phase is complete, the CDA technique can be conducted using pen and paper.

Summary flowchart

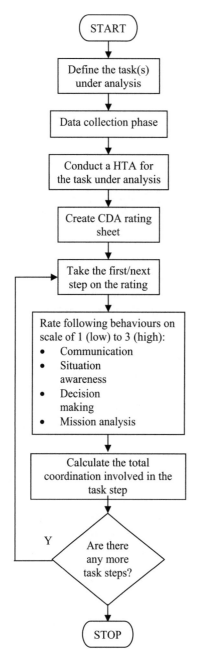

CUD – Comms Usage Diagram

Background and applications

Comms Usage Diagram (CUD) (Watts and Monk, 2000) is used to describe collaborative activity between teams of agents dispersed across different geographical locations. A CUD output describes the order of communications during the scenario under analysis, how and why communications between agents occur, which technology is involved in the communication, and the advantages and disadvantages of the technology used. The CUD technique was originally developed and applied in the telecommunications domain, whereby the technique was used to analyse 'telemedical consultation' involving a medical practitioner offering advice regarding a medical ailment from a different location to the advice seeker (Watts and Monk, 2000). The technique has more recently been used by the authors in the analysis of C4i activity in a number of domains, including energy distribution, naval warfare, fire services, air traffic control, military, rail and aviation domains (see Chapters 4–6)). A CUD analysis is typically based upon observational data of the task or scenario under analysis, although talk through analysis and interviews are also used (Watts and Monk, 2000).

Domain of application

Although the technique was originally developed for use in the medical domain, it is generic and can be applied in any domain that involves distributed collaboration.

Procedure and advice

Step 1: Define the task or scenario under analysis The first step in a CUD analysis is to clearly define the task or scenario under analysis. It may be useful to conduct a HTA of the task under analysis for this purpose. A clear definition of the task under analysis allows the analyst(s) to prepare for the data collection phase.

Step 2: Data collection Next, the analyst(s) should collect specific data regarding the task or scenario under analysis. Watts and Monk (2000) recommend that interviews, observations and task talk-through should be used to collect the data. Specific data regarding the personnel involved, activity, task steps, communication between personnel, technology used and geographical location should be collected.

Step 3: Complete Initial Comms report Following the data collection phase, the raw data obtained should be put into a report form. According to Watts and Monk (2000) the report should include the location of the technology used, the purpose of the technology, the advantages and disadvantages of using such technology, and graphical account of a typical consultation session. The report should then be made available to all personnel involved for evaluation and reiteration purposes.

Step 4: Construct CUD output table The graphical account developed in step 3 forms the basis for the CUD output. First of all, the analyst should construct an appropriate CUD template, containing a column for each agent involved in the scenario, and columns for the technology used, the advantages and disadvantages associated with the technology used and the recommended technology. An example CUD template is presented in Figure 3.11. The CUD output contains a description of the task activity at each geographical location and the collaboration between personnel at each location. Directional arrows (that is, from/to) should then be used to represent the communications between personnel at different locations. The comms column specifies the technology used in the communication and the effects column lists any advantages and disadvantages associated with the technology in question. This is based upon the initial observation of the scenario and also analyst subjective judgement. Finally, the recommended comms column is used to provide a recommendation of the most suitable technology available for the comms in question.

Step 5: Calculate communications technology figures In order to summarise the CUD analysis, the frequency of communications made by each agent and the communications technology used is calculated. An example CUD summary table is presented in Table 3.17.

Advantages

- The CUD output is particularly useful, offering a description of the task under analysis, and also a description of collaborative activity involved, including the order of activity, the communications between agents, the personnel involved, the technology used and its associated advantages and disadvantages, and recommendations regarding the technology used.
- The output of a CUD analysis is particularly useful for highlighting communication flaws in a particular network of agents.

- The CUD technique is particularly suited to the analysis of teamwork and C4i activity.
- The CUD technique is also flexible, and could potentially be modified to make it more comprehensive. Factors such as time, error and workload could be incorporated into a CUD analysis, ensuring a much more exhaustive analysis.
- Although the CUD technique was developed and originally used in the medical domain, it is a generic technique and could potentially be applied in any domain involving distributed collaboration.
- The technique is easy to use and quick to learn.

Disadvantages

- For large, complex tasks involving many agents, conducting a CUD analysis may become a lengthy and laborious process.
- In its present usage, the CUD analysis only defines the communications technology used at the source of the communication in question (that is, the comms resource at the other end of the communication in question is not defined).
- The initial data collection phase of the CUD technique is time consuming and labour intensive.
- No validity or reliability data are available for the technique.
- Application of the CUD technique appears to be limited.
- No guidelines are offered to the analyst in analysing the technology used in the communications under analysis. Typically this is based upon the analyst's subjective judgement. This may affect the reliability of the technique.

Example

The following example is a CUD analysis of an energy distribution task. The task involved the return from isolation of the Brimsdown-Tottenham-Waltham Cross Circuit (Salmon et al., 2004b; 2004c). The data collection phase involved two observers. The first observer was situated at the National Grid Transco (NGT) National Operations Centre (NOC) in Warwick and observed the activity of the NOC control room operator. The second observer was situated at the Tottenham substation and observed the activity of the senior authorised person (SAP) and authorised person (AP) who completed work required to return the circuit from isolation. The CUD analysis used observational data and a HTA of the task as its primary inputs. An extract of the CUD analysis for the energy distribution task is presented in Figure 3.12. In order to summarise the CUD analysis, comms frequency figures were calculated. The CUD summary results are presented in Table 3.17.

Related methods

During the data collection phase, a number of different techniques may be used, such as observational analysis, interviews and talk-through type analysis. It is also useful to conduct a HTA of the task under analysis prior to constructing the CUD. The CUD technique has also recently been integrated with a number of other techniques (HTA, observation, co-ordination demands analysis, social network analysis, operator sequence diagrams and propositional networks) to form the event analysis of systemic teamwork (EAST) methodology, which has been used to analyse C4i activity.

Approximate training and application times

The training time associated with CUD is very low, assuming that the practitioner was already proficient in data collection techniques such as interviews and observational analysis. In a recent training programme for the EAST methodology, the CUD training session lasted under one hour. The application time associated with

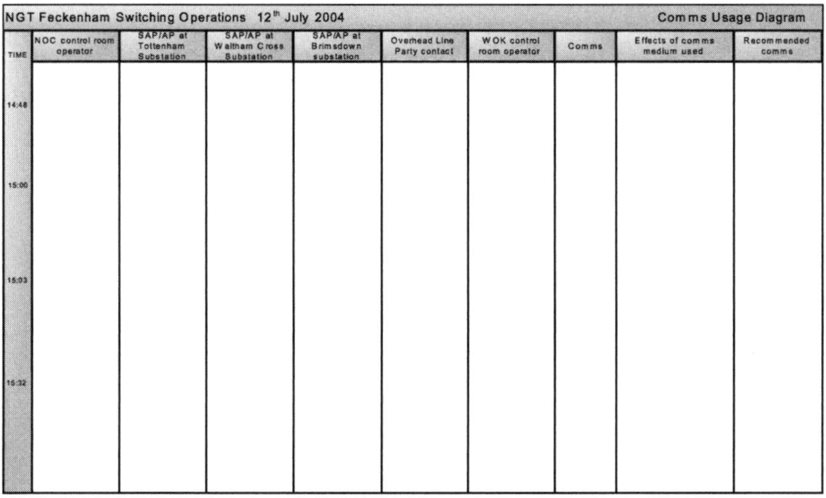

Figure 3.11 CUD template

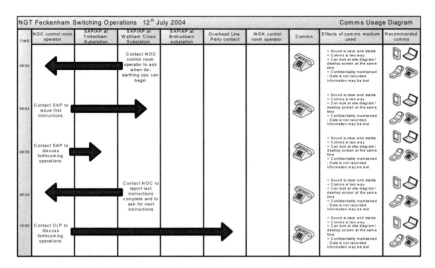

Figure 3.12 Comms usage diagram for an energy distribution task (Salmon et al., 2004)

the CUD technique is also typically low to medium, depending upon task size and complexity. In the analysis of energy distribution scenarios (Salmon et al., 2004a, b, c) the application time for CUD ranged between 1 and 4 hours, depending upon the size and complexity of the scenario under analysis.

Reliability and validity
No data regarding the reliability and validity of the technique are available.

Tools needed
A CUD analysis can be conducted using pen and paper only. However, it is recommended that video and audio recording equipment are used during the data collection phase. The CUD can be constructed using Microsoft Visio.

Summary flowchart

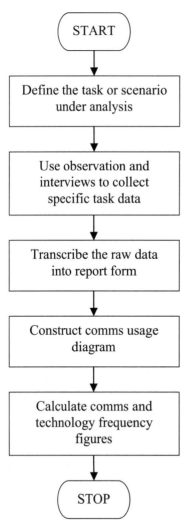

Social Network Analysis

Background and applications
Social Network Analysis (SNA) is used to analyse and represent the relationships between teams or agents within a social network. A social network is defined as a set or team of agents that possess relationships with one another (Driskell and Mullen, 2005). SNA is based upon the notion that the relationship between agents within a social network has a significant effect upon the actions performed and also the performance achieved by the network. SNA uses both graphical and mathematical procedures to represent social networks. Typically, centrality measures are calculated for each agent (for example, degree, betweenness and closeness) and the overall network density is calculated. This allows the identification of the key agents within the network and also the classification of the network structure. The technique has previously been used for the analysis of networks in a number of areas, including military C4ISR architectures (Dekker, 2002), C4i activity in the fire service (Baber et al., 2004), energy distribution (Salmon et al., 2004b; 2004c), air traffic control, rail (Walker et al., 2005; 2006b), naval warfare (Stewart et al., 2004) and military aviation domains, counselling and computer-mediated communications (Dekker, 2002).

Domain of application
Generic.

Procedure and advice

Step 1: Define network or group The first step in a SNA involves defining the network or group of networks that are to be analysed. Once the overall network type is specified, further social networks within the network type specified should be defined. For the analysis of networks, the authors identified a number of different networks that could be analysed. These were networks within the emergency services, rail, military, aviation, air traffic control, and civil energy distribution.

Step 2: Define scenarios Typically, networks are analysed over a number of different scenarios. Once the type of network under analysis has been defined, the scenario(s) within which they will be analysed should be defined. For a thorough analysis, a number of different scenarios should be analysed. The scenarios used should be representative of all C4i activity within the network under analysis. In the analysis of naval warfare C4i (Stewart et al., 2004), the following scenarios were analysed:

- air threat scenario
- surface threat scenario
- subsurface threat scenario.

Step 3: Data collection Once the network and scenario(s) under analysis are defined clearly, the data collection phase can begin. The data collection phase involves the collection of specific data on the relationship (for example communications and activity) between the agents during the scenario. Typical HF data collection techniques should be used in this process, such as observational analysis, interviews and questionnaires. Typically agent activity and the frequency, direction and content of any communications between agents in the network are recorded.

Step 4: Construct agent association matrix Once sufficient data regarding the scenario under analysis is collected, an agent association matrix should be constructed. The matrix represents the frequency of associations between each agent in the network.

Step 5: Construct social network diagram Once the matrix of association is completed, the social network diagram should be created. The social network depicts each agent in the network and the communications between them. The communications are represented by directional arrows, and the frequency of communications is also presented. An example social network diagram is presented in Figure 3.13.

Step 6: Calculate agent centrality Agent centrality is calculated in order to determine the central or key agent(s) within the network. There are a number of different centrality calculations that can be made. For example, agent centrality can be calculated using Bavelas-Leavitt's index. The mean centrality + standard deviation can then be used to define key agents within the network. Those agents who posses a centrality figure that exceeds the mean + standard deviation figure are defined as the key agents within the network.

Step 7: Calculate sociometric status The sociometric status of each agent refers to the number of communications received and emitted, relative to the number of nodes in the network. The mean sociometric status + standard deviation can also be used to define key agents within the network. Those agents who posses a sociometric status figure that exceeds the mean + standard deviation figure can be defined as the key agents within the network.

Step 8: Calculate network density Network density is equal to the total number of links between the agents in the network divided by the total number of possible links. Low network density figures are indicative of a well distributed network of agents. High density figures indicate a network that is not well distributed.

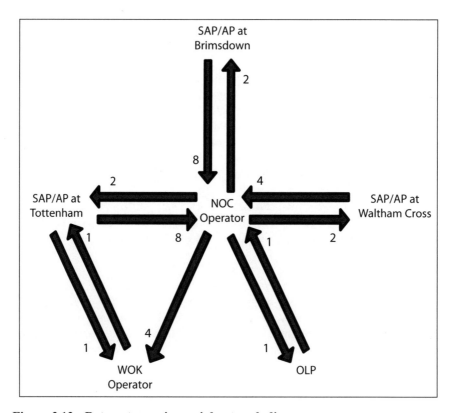

Figure 3.13 Return to service social network diagram

Advantages

- SNA can be used to determine the importance of different agents within a social network and also to classify the network type.
- The SNA offers a comprehensive analysis of the network in question. The key agents within the network are specified, as are the frequency and direction of communications within the network. Further classifications include network type and network density. There are also additional analyses that can be calculated, such as betweenness, closeness and distance calculations.
- Networks can be classified according to their structure. This is particularly useful when analysing networks across different domains.
- SNA is particularly suited to the analysis of C4i scenarios.
- The technique has been used extensively in the past for the analysis of various social networks.
- The technique is simple to learn and easy to use.
- SNA is a very quick technique to apply. The provision of the AGNA SNA software package also reduces application time further.
- SNA is a generic technique that could potentially be applied in any domain.

Disadvantages

- For large, complex networks, it may be difficult to conduct a SNA. Application time is a function of network size, and large networks may incur lengthy application times.
- The data collection phase involved in a SNA is also resource intensive.
- Some knowledge of mathematical techniques is required.
- It is difficult to collect comprehensive data for a SNA. For example, a dispersed network of ten agents would require at least 10 observers in order to accurately capture the communications made between all agents.

Example

The following example is taken from an EAST analysis of a civil energy distribution scenario (Salmon et al., 2004b; 2004c). The scenario involved the return to service of the Brimsdown-Tottenham-Waltham Cross Circuit and took place at Brimsdown, Tottenham, Waltham Cross substations and the National Grid Transco (NGT) Network Operations Centre (NOC) in Warwick. The agents involved in the scenario are presented in Table 3.18.

From the list of agents identified for the scenario, a matrix of association can be constructed. This matrix shows whether or not an agent within the system can be associated with any other agent, specifically through frequency of communications. The association matrix for the switching scenario is presented in Table 3.19.

Finally, a social network diagram is created. The social network diagram illustrates the proposed association between agents involved in the scenario. The numbers associated with the links between the agents in the system indicate the strength of association. The strength of association is defined by the number of occasions on which agents exchanged information. The direction of association is represented by directional arrows. The social network diagram is presented in Figure 3.13.

Table 3.18 Agents involved in the return to service scenario

Role of agent A	NOC control room operator
Role of agent B	SAP/AP at Tottenham substation
Role of agent C	SAP/AP at Waltham Cross substation
Role of agent D	SAP/AP at Brimsdown substation
Role of agent E	Overhead line party contact
Role of agent F	WOK control room operator

Table 3.19 Agent association matrix

	A	B	C	D	E	F
A	0	2	2	2	1	4
B	8	0	0	0	0	1
C	4	0	0	0	0	0
D	8	0	0	0	0	0
E	1	0	0	0	0	0
F	0	1	0	0	0	0

Table 3.20 Agent centrality (B-L Centrality)

Agent	B-L Centrality
NOC operator	4.72
SAP/AP at Tottenham	3.25
SAP/AP at Waltham Cross	2.73
SAP/AP at Brimsdown	2.73
Overhead line party	2.73
WOK operator	2.6

Event Analysis of Systemic Team-work 101

There are a number of ways to analyse social networks. In this case, agent centrality, sociometric status and network density were calculated. Agent centrality was calculated using Bavelas-Leavitt's index. Table 3.20 shows the Centrality for the agents in this incident. The mean centrality was calculated as 3.13. A notion of 'key' agents can be defined using the mean + 1 standard deviation (that is 3.13 + 0.74 = 3.87). Using this rule, the B-L centrality calculation indicates that the NOC operator and the SAP/AP at Tottenham substation are the key agents in the network.

Table 3.21 shows the Sociometric Status for each agent involved in the scenario. From the calculation, a mean status of 2.26 (±3.82) was found. The value of mean + one standard deviation, that is, 2.26 + 3.82 = 6.08, is used to define 'key' agents in this network. Again, the sociometric status analysis indicates that the NOC operator is the key agent within the network.

An overall measure of network density was also derived by dividing the links actually present in the scenario, by all of the available links. For the Tottenham scenario, the overall network density is calculated as 0.2 (6 links present divided by 30 possible links). This figure is suggestive of a well distributed (and therefore less dense) network of agents.

Related methods
In terms of rating relationships in networks the SNA technique appears to be unique. SNA also uses mathematical techniques for the analysis of the networks involved. During the data collection phase, techniques such as observational study, interviews and questionnaires are typically used.

Approximate training and application times
The associated training time for the SNA methodology is low. Although some knowledge of mathematical analysis is required, the basic SNA procedure is a simple one. It is estimated that the technique could be trained to HF practitioners within 1 hour. The application time for the SNA technique is also minimal. The provision of AGNA SNA software package reduces the application time even further. In a SNA of an energy distribution switching scenario (involving a network of four agents) the application time was approximately 2 hours. However, the data collection phase

Table 3.21 Agent sociometric status

Agent	Sociometric status
NOC operator	6.4
SAP/AP at Tottenham	2.4
SAP/AP at Waltham Cross	1.2
SAP/AP at Brimsdown	2.0
Overhead line party	0.4
WOK operator	1.2

involved adds to the overall application time, and this is typically lengthy. Also, the application time is dependent upon the size and complexity of the network under analysis, larger more complex networks may incur greater application times.

Reliability and validity
No data regarding the reliability and validity of the SNA technique are available.

Tools needed
SNA can be conducted using pen and paper, once the data collection phase is complete. However, there are various SNA software packages that can be used that automate the SNA procedure. For example, based upon an association matrix as input, the AGNA software package constructs the social network and also performs various statistical analyses of the network in question. Further software packages are available, such as UCINET (Borgatti, Everett and Freeman, 2002) and STRUCTURE (Burt, 1991). The tools required during the data collection phase for a SNA would be dependent upon the type of data collection techniques used. Observational analysis, interviews and questionnaires would normally require visual and audio recording equipment (video cameras, minidisc recorder, PC).

Summary flowchart

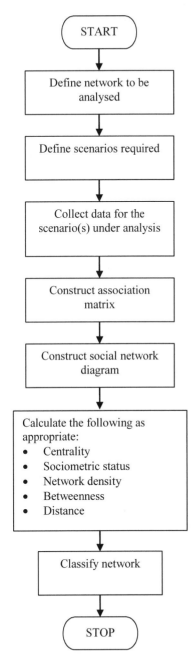

Operation sequence diagrams

Background and applications
Operation sequence diagrams (OSD) are used to graphically describe the activity and interaction between teams of agents within a network. According to Kirwan and Ainsworth (1992), the original purpose of OSD analysis was to represent complex, multi-person tasks. The output of an OSD graphically depicts the task process, including the tasks performed and the interaction between operators over time, using standardised symbols. There are various forms of OSDs, ranging from a simple flow diagram representing task order, to more complex OSD, which account for team interaction and communication. OSDs have been applied in a number of domains, including the fire service (Baber et al., 2004), naval warfare (Stewart et al., 2004), aviation (Stewart et al., 2004), energy distribution (Salmon et al., 2004b; 2004c), air traffic control (Walker et al., 2005) and rail (Walker et al., 2006b) domains.

Domain of application
The technique was originally used in the nuclear power and chemical process industries. However, the technique is generic and can be applied in any domain.

Procedure and advice

Step 1: Define the task(s) under analysis The first step in an OSD analysis is to define the task(s) or scenario(s) under analysis. The task(s) or scenario(s) should be defined clearly, including the activity and agents involved.

Step 2: Data collection In order to construct an OSD, the analyst(s) must obtain sufficient data regarding the task or scenario under analysis. It is recommended that the analyst(s) use various forms of data collection in this phase. Observational study should be used to observe the task (or similar types of task) under analysis. Interviews with personnel involved in the task (or similar tasks) should also be conducted. The type and amount of data collected in step 2 is dependent upon the analysis requirements. The more exhaustive the analysis is intended to be, the more data collection techniques should be used.

Step 3: Describe the task or scenario using HTA Once the data collection phase is completed, a detailed task analysis should be conducted for the scenario under analysis. The type of task analysis is determined by the analyst(s), and in some cases, a task list will suffice. However, it is recommended that a HTA is conducted.

Step 4: Construct the OSD diagram Once the task has been described adequately, the construction of the OSD can begin. The OSD template should include the title of the task step, a timeline, and a row for each agent involved in the task. An OSD template is presented in Figure 3.14. In order to construct the OSD, it is recommended that the analyst walks through the HTA of the task under analysis, constructing the OSD in conjunction with this. An OSD symbol glossary is presented in Figure 3.3. The symbols involved in a particular task step should be linked by directional arrows, in order to represent the flow of activity during the task. Each symbol in the OSD should contain the corresponding task step number from the HTA of the task under analysis. The artefacts used during the communications should also be annotated onto the OSD.

Step 5: Overlay additional analyses results One of the endearing features of the OSD technique is that additional analysis results can easily be added to the OSD. According to the analysis requirements, additional task features can also be annotated onto the OSD. For example, Baber et al. (2004) added total co-ordination values for teamwork task steps (from an initial CDA analysis) onto the OSD for a fire service task.

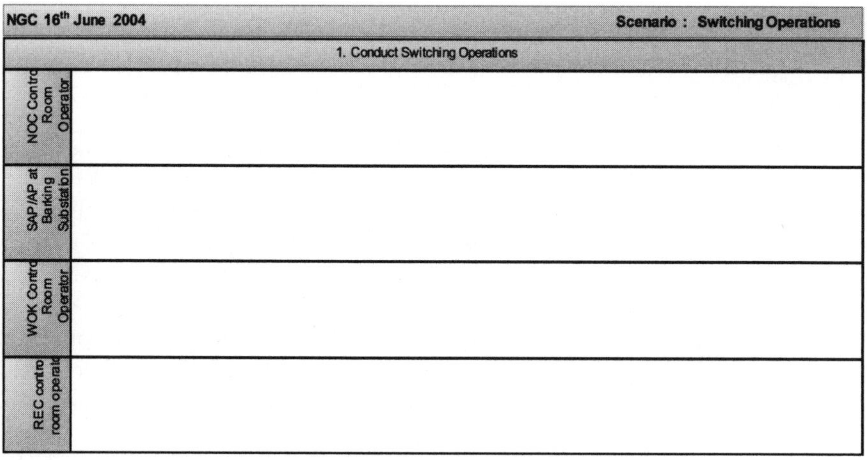

Figure 3.14 Example OSD template

Step 6: Calculate operation loading figures   From the OSD, operational loading figures are calculated for each agent involved in the scenario under analysis. Operational loading figures are calculated for each OSD operator or symbol used, for example operation, receive, delay, decision, transport, and combined operations. The operational loading figures refer to the frequency in which each agent was involved in the operation in question during the scenario.

Advantages

- The OSD provides an exhaustive analysis of the task in question. The flow of the task is represented in terms of activity and information, the type of activity and the agents involved is specified, a timeline of the activity, the communications between agents involved in the task, the technology used and also a rating of total co-ordination for each teamwork activity is also provided. The technique's flexibility also permits the analyst(s) to add further analysis outputs onto the OSD, adding to its exhaustiveness.
- An OSD is particularly useful for analysing and representing distributed teamwork or collaborated activity.
- OSDs are useful for demonstrating the relationship between tasks, technology and team members.
- High face validity (Kirwan and Ainsworth, 1992).
- OSDs have been used extensively in the past and have been applied in a variety of domains involving team based activity, including the emergency services (Baber et al., 2004), energy distribution (Salmon et al., 2004b; 2004c), rail (Walker et al., 2006b), air traffic control (ATC) (Walker et al., 2005) and naval operations (Stewart et al., 2004).
- A number of different analyses can be overlaid onto an OSD of a particular task. For example, Baber et al. (2004) add the corresponding HTA task step numbers and co-ordination demands analysis results to OSDs of C4i activity.
- The OSD technique is very flexible and can be modified to suit the analysis needs.
- The WESTT software package can be used to automate a large portion of the OSD procedure.

Disadvantages

- The application time for an OSD analysis is lengthy. Constructing an OSD for large, complex tasks can be extremely time consuming and the initial data collection adds further time to the analysis.
- The construction of large, complex OSDs is also quite difficult.
- OSDs can become cluttered and confusing (Kirwan and Ainsworth, 1992).
- The output of OSDs can become large and unwieldy.
- The present OSD symbols are limited for certain applications (for example, C4i scenarios).
- The reliability of the technique is questionable. Different analysts may interpret the OSD symbols differently.

Example

The following example is an extract of an OSD of an energy distribution scenario (Salmon et al, 2004b; 2004c). The task involved the switching out of three circuits (SGT5 and SGT1A and 1B) at Barking 275Kv, 132Kv and 33Kv Substations. Circuit SGT5 was being switched out for the installation of a new transformer for the nearby channel tunnel rail link and SGT1A and 1B were being switched out for substation maintenance. Observational data from Barking substation and the NOC control room was used to conduct a HTA of the switching scenario. A HTA was then created, which acted as the primary input into the OSD diagram. Total co-ordination values for each teamwork task step are also annotated onto the OSD. The glossary for the OSD has been presented in an earlier section (see Figure 3.3). An extract of the HTA for the corresponding energy distribution task follows. The corresponding extract of the OSD was also presented in Figure 3.4. The operational loading figures were presented in Table 3.8.

The operational loading analysis indicates that the senior authorised person (SAP) at Barking substation has the highest loading in terms of operations, transport, and delay whilst the network operations centre (NOC) operator has the highest loading in terms of receipt of information. This provides an indication of the nature of the roles involved in the scenario. The NOC operator's role is one of information distribution (giving and receiving) indicated by the high receive operator loading, whilst the majority of the work is conducted by the SAP at Barking, indicated by the high operation and transport loading figures.

Extract of HTA for the NGT switching scenario (Source: Salmon et al., 2004b)

0. Co-ordinate *and* carry out switching operations on circuits SGT5. SGT1A and 1B at Bark s/s (Plan 0. Do 1 then 2 then 3, EXIT)

1. Prepare for *switching* operations (Plan 1. Do 1.1, then 1.2, then 1.3, then 1.4, then 1.5, then 1.6, then 1.7, then 1.8, then 1.9, then 1.10 EXIT)

    1.1. Agree SSC (Plan 1.1. Do 1.1.1, then 1.1.2, then 1.1.3, then 1.1.4, then 1.1.5, EXIT)

        1.1.1. (WOK) Use phone to Contact NOC
        1.1.2. (WOK + NOC) Exchange identities
        1.1.3. (WOK + NOC) Agree SSC documentation
        1.1.4. (WOK+NOC) Agree SSC and time (Plan 1.1.4. Do 1.1.4.1, then 1.1.4.2, EXIT)

            1.1.4.1. (NOC) Agree SSC with WOK
            1.1.4.2. (NOC) Agree time with WOK

        1.1.5. (NOC) Record and enter details (Plan 1.1.5. Do 1.1.5.1, then 1.1.5.2, EXIT)

            1.1.5.1. Record details on log sheet
            1.1.5.2. Enter details into worksafe

    1.2. (NOC) Request remote isolation (Plan 1.2. Do 1.2.1, then 1.2.2, then 1.2.3, then 1.2.4, EXIT)

        1.2.1. (NOC) Ask WOK for isolators to be opened remotely

1.2.2. (WOK) Perform remote isolation
1.2.3. (NOC) Check Barking s/s screen
1.2.4. (WOK + NOC) End communications
1.3. Gather information on outage at transformer 5 at Bark s/s
(Plan 1.3. Do 1.3.1, then 1.3.2, then 1.3.3, then 1.3.4, EXIT)
1.3.1. (NOC) Use phone to contact SAP at Bark

Related methods
Various types of OSD exist, including temporal operational sequence diagrams, partitioned operational sequence diagrams and spatial operational sequence diagrams (Kirwan and Ainsworth, 1992). During the OSD data collection phase, techniques such as observational study and interviews are typically used. Task analysis techniques such as HTA are also used during the construction of the OSD. Timeline analysis may also be used in order to construct an appropriate timeline for the task or scenario under analysis. Additional analyses results can also be annotated onto an OSD, such as CDA and comms usage diagram.

Approximate training and application times
The training time associated with the OSD methodology is low. However, the typical application time for the technique is high. The construction of an OSD can be a very time consuming process, involving many re-iterations. A typical OSD normally can take 12 + hours to construct.

Reliability and validity
According to Kirwan and Ainsworth (1992), OSD techniques possess a high degree of face validity. The intra analyst reliability of the technique may be suspect, as different analysts may interpret the OSD symbols differently.

Tools needed
When conducting an OSD analysis, pen and paper could be sufficient. However, to ensure that data collection is comprehensive, it is recommended that video or audio recording devices are used in conjunction with the pen and paper. For the construction of the OSD, it is recommended that a suitable drawing package, such as Microsoft Visio is used. The WESTT software package can also be used to automate a large portion of the OSD procedure. WESTT constructs the OSD based upon an input of the HTA for the scenario under analysis.

Summary flowchart

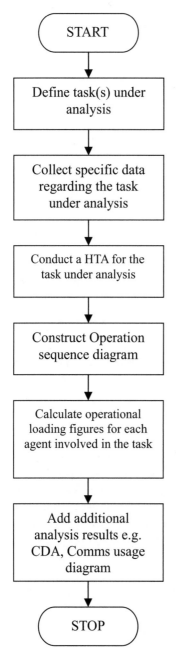

Critical decision method

Background and applications
The Critical Decision Method is a semi-structured interview technique that uses cognitive probes in order to elicit information regarding expert decision-making. According to the authors, the technique can serve to provide knowledge engineering for expert system development, identify training requirements, generate training materials and evaluate the task performance impact of expert systems (Klein, Calderwood and MacGregor, 1989). The technique is an extension of the Critical Incident Technique (Flanagan, 1954) and was developed in order to study the naturalistic decision-making strategies of experienced personnel. The CDM is perhaps the most commonly used cognitive task analysis technique and has been applied in a number of domains, including the fire service (Baber et al., 2004), military and paramedics (Klein, Calderwood and MacGregor, 1989), air traffic control (Walker et al., 2005), civil energy distribution (Salmon et al., 2004b; 2004c), naval warfare (Stewart et al., 2004), rail (Walker et al., 2006b) and even white water rafting (O'Hare et al., 2000).

Domain of application
Generic.

Procedure and advice

Step 1: Select the incident to be analysed The first part of a CDM analysis is to select the incident that is to be analysed. Depending upon the purpose of the analysis, the type of incident may already be selected. CDM normally focuses on non-routine incidents, such as emergency incidents, or highly challenging incidents. If the scenario under analysis is not already specified, the analyst(s) may identify an appropriate incident via interview with an appropriate SME, by asking them to describe a recent highly challenging (that is, high workload) or non routine incident in which they were involved. The interviewee involved in the CDM analysis should be the primary decision maker in the chosen incident.

Step 2: Select CDM probes   The CDM technique works by probing a SME using specific probes designed to elicit pertinent information regarding the decision making process during key points in the incident under analysis. In order to ensure that the output is compliant with the original aims of the analysis, an appropriate set of CDM probes should be defined prior to the analysis. The probes used are dependent upon the aims of the analysis and the domain in which the incident is embedded. Alternatively, if there are no adequate probes available, the analyst(s) can develop novel probes based upon the analysis needs. A set of CDM probes from O'Hare et al. (2000) is presented in Table 3.9.

Step 3: Select appropriate participant   Once the scenario under analysis and the probes to be used are defined, the analyst(s) should proceed to identify an appropriate SME or set of SMEs. Typically, operators of the system under analysis are used.

Step 4: Gather and record account of the incident   The CDM procedure can be applied to an incident observed by the analyst or to a retrospective incident described by the participant. If the CDM is based upon an observed incident, then this step involves firstly observing the incident and then recording an account of the incident. Otherwise, the incident can be described retrospectively from the participant's memory. The analyst should ask the SME for a description of the incident in question, from its starting point to its end point.

Step 5: Construct incident timeline   The next step in the CDM analysis is to construct a timeline of the incident described in step 4. The aim of this is to give the analyst(s) a clear picture of the incident and its associated events, including when each event occurred and what the duration of each event was. According to Klein, Calderwood and MacGregor (1989) the events included in the timeline should encompass any physical events, such as alarms sounding, and also 'mental' events, such as the thoughts and perceptions of the interviewee during the incident.

Step 6: Define scenario phases   Once the analyst has a clear understanding of the incident under analysis, the incident should be divided into key phases or decision points. This should be done in conjunction with the SME. Normally, the incident is divided into four or five key phases.

Step 7: Use CDM probes to query participant decision making   For each incident phase, the analyst should probe the SME using the CDM probes selected during step 2 of the procedure. The probes are used in an unstructured interview format in order to gather pertinent information regarding the SMEs decision making during each incident phase. The interview should be recorded using an audio recording device such as a mini-disc recorder.

Step 8: Transcribe interview data   Once the interview is complete, the data should be transcribed accordingly.

Step 9: Construct CDM tables   Finally, a CDM output table for each scenario phase should be constructed. This involves simply presenting the CDM probes and the associated SME answers in an output table (see Tables 3.10–3.13).

Advantages

- The CDM analysis can be used to elicit specific information regarding the decision making strategies used in complex systems.
- The CDM output can be used to construct propositional networks that describe the knowledge or SA objects required during the scenario under analysis.
- The technique is normally quick in application.
- Once familiar with the technique, CDM is relatively easy to apply.
- The CDM is a popular procedure and has been applied in a number of domains.

Disadvantages

- The reliability of such a technique is questionable. Klein and Armstrong (2005) suggest that methods that analyse retrospective incidents are associated with concerns of data reliability, due to evidence of memory degradation.
- The data obtained is highly dependent upon the skill of the analyst conducting the CDM interview and also the quality of the participant used.
- A high level of expertise and training is required in order to use the CDM to its maximum effect (Klein and Armstrong, 2005).
- The CDM relies upon interviewee verbal reports in order to reconstruct incidents. How far a verbal report accurately represents the cognitive processes of the decision maker is questionable. Facts could be easily misrepresented by the participants involved.
- It is often difficult to gain sufficient access to appropriate SMEs in order to conduct a CDM analysis.

Example
A worked example of a CDM analysis from the civil energy distribution domain (Salmon et al., 2004b; 2004c) has previously been given in section 3.2.1.2.

Related methods
The CDM is an extension of the critical incident technique (Flanagan, 1954). The CDM is also closely related to other cognitive task analysis (CTA) techniques, in that it uses probes to elicit data regarding task performance from participants. Other similar CTA techniques include ACTA (Militello and Hutton, 2000) and cognitive walkthrough analysis (Polson et al., 1992).

Approximate training and application times
Klein and Armstrong (2005) report that the training time associated with the CDM would be high. Experience in interviews with SMEs is required, and also a grasp of cognitive psychology. The application time for the CDM is medium. The CDM interview takes between 1 – 2 hours, and the transcription process takes approximately 1 – 2 hours.

Reliability and validity
Both intra and inter analyst reliability of the CDM approach is questionable. It is apparent that such an approach may elicit different data from similar incidents when applied by different analysts on separate participants. Klein and Armstrong (2005) suggest that there are also concerns associated with the reliability of the CDM due to evidence of memory degradation.

Tools needed
When conducting a CDM analysis, pen and paper could be sufficient. However, to ensure that data collection is comprehensive, it is recommended that video or audio recording equipment is used. A set of CDM probes, such as those presented in Table 3.9 are also required. The type of probes used is dependent upon the focus of the analysis.

Summary flowchart

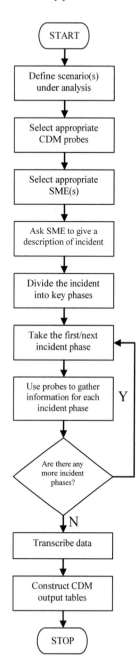

Propositional networks

Background and applications
Propositional networks are used to identify the knowledge objects related to a particular task or scenario, and also the links between each of the knowledge objects identified. According to Baber and Stanton (2004), the concept of representing 'knowledge' in the form of a network has been subjected to major discussion within cognitive psychology since the 1970s. Propositional networks consist of a set of nodes that represent knowledge, sources of information, agents, and artefacts that are linked through specific causal paths. Thus the propositional network offers a way of presenting the 'ideal' collection of knowledge required during the scenario in question. Networks are constructed from an initial critical decision method analysis of the scenario in question. A simple content analysis is used to identify the knowledge objects for each scenario phase as identified by the CDM analysis. A propositional network is then constructed for each phase identified by the CDM analysis, comprised of the knowledge objects and the links between them. Propositional networks have been used to represent knowledge and shared situation awareness as part of the EAST methodology (Baber et al., 2004), which has been used to analyse C4i scenarios in the civil energy distribution domain (Salmon et al., 2004b; 2004c), the rail domain (Walker et al., 2006b), the air traffic control domain (Walker et al., 2005), the naval warfare domain (Stewart et al., 2004) and the emergency services domain (Baber et al., 2004).

Domain of application
Generic.

Procedure and advice

Step 1: Define scenario The first step in a propositional network analysis is to define the scenario under analysis. The scenario in question should be defined clearly. This allows the analyst(s) to determine the data collection procedure that follows and also the appropriate SMEs required for the CDM phase of the analysis.

Step 2: Conduct a HTA for the scenario Once the scenario has been clearly defined, the next step involves describing the scenario using HTA. A number of data collection techniques may be used in order to gather the information required for the HTA, such as interviews with SMEs and observations of the task under analysis.

Step 3: Conduct a CDM analysis The propositional networks are based upon a CDM analysis of the scenario in question. The CDM analysis should be conducted using appropriate SMEs (see the previous section for a full description of the CDM procedure). The CDM involves dividing the scenario under analysis into a number of key phases and then probing the SME using pre-defined 'cognitive' probes, designed to determine pertinent features associated with decision making during each scenario phase.

Step 4: Conduct content analysis Once the CDM data are collected, a simple content analysis should be conducted for each phase identified during the CDM analysis. In order to convert the CDM tables into propositions, a content analysis is performed. In the first stage, this simply means separating all content words from any function words. Consider the following quote:

> There is currently a serious water shortage within the internally displaced person (IDP) camps. We need to identify water sources, such as streams and local suppliers, means of distribution and transport, including routes, and also storage facilities

For example, from the above quote entry we would identify the words in italic as the knowledge elements from the CDM data, as listed:

- water
- shortage
- IDP
- IDP camps
- sources
- streams
- local Suppliers
- distribution
- transport
- routes
- storage.

Step 5: Define links between knowledge objects Once the knowledge elements for each scenario phase have been identified, the next step involves defining the links between the knowledge elements in each phase. The following taxonomy is used:

- has
- is
- are
- causes

- knows
- needs
- requires
- prevents.

For those knowledge elements that are linked during the scenario, the type of link should be defined using the taxonomy above.

Step 6: Construct propositional networks The final step is to construct the propositional network diagrams for each scenario phase. A propositional network diagram should be constructed for the overall scenario (that is including all knowledge elements) and then separate propositional network diagrams should be constructed for each phase, with the knowledge elements required highlighted. Further coding of the knowledge elements may also be used, for example shared knowledge elements can be striped, and inactive knowledge elements that have been used in previous scenario phases are typically shaded.

Working through the list of knowledge elements leads to a set of propositions. These are checked to ensure that duplication is minimised and then used to construct the propositional network, as shown in Figure 3.15.

In order to specify the knowledge objects for each phase, the analyst simply takes the CDM output for each phase and using content analysis, identifies the required knowledge elements. Knowledge elements include any knowledge, information, agents and artefacts identified by the CDM analysis. A list of knowledge elements should be made for each scenario phase.

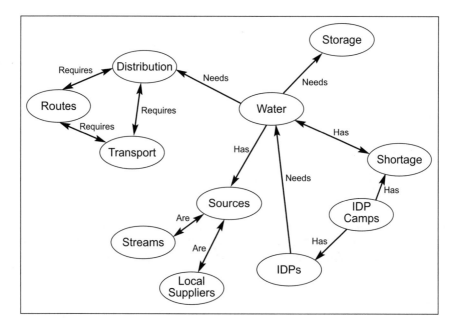

Figure 3.15 Example propositional network

Advantages

- The output represents the ideal collection of knowledge required for performance during the scenario under analysis.
- The knowledge objects are defined for each phase of the scenario under analysis, and the links between the knowledge objects are also specified.
- The technique is easy to learn and use.
- The technique is also quick in its application.
- Propositional networks are ideal for analysing teamwork and representing shared situation awareness during a particular scenario.

Disadvantages

- The initial HTA and CDM analysis add considerable time to the associated application time.
- Inter and intra analyst reliability of the technique is questionable.
- A propositional network analysis is reliant upon acceptable CDM data.
- It may be difficult to gather appropriate SMEs for the CDM part of the analysis.

Example
This example is taken from an analysis of a switching scenario drawn from the civil energy distribution domain (Salmon et al., 2004b). The propositional networks have already been presented in Figures 3.5–3.9. These present the knowledge objects (darker shading) identified from the corresponding CDM output for that phase. The CDM outputs are presented in Tables 3.10–3.13. The propositional network consists of a set of nodes that represent sources of information, agents, and objects etc. that are linked through specific causal paths. From this network, it is possible to identify required information and possible options relevant to this incident. The concept behind using a propositional network in this manner is that it represents the 'ideal' collection of knowledge for the scenario. As the incident unfolds, so participants will have access to more of this knowledge (either through communication with other agents or through recognising changes in the incident status). Consequently, within this propositional network, Situation Awareness can be represented as the change in weighting of links. Propositional networks were developed for the overall scenario and also the incident phases identified during the CDM analysis. The propositional networks indicate which of the knowledge objects are active (that is agents are using them) during each incident phase. The lighter nodes in the propositional networks represent unactivated knowledge objects (that is knowledge is available but is not required nor is it being used). The dark nodes represent active (or currently being used) knowledge objects.

Related methods
Propositional networks require an initial CDM analysis as an input. A HTA is also typically conducted prior to the propositional network analysis.

Approximate training and application times
The propositional network methodology requires only minimal training. In a recent HF methods training session, the training time for the propositional network technique was approximately one hour. However, the analyst should be competent in the HTA and CDM procedure in order to conduct the analysis properly. The application time for propositional networks alone is high, as it involves a content analysis (on CDM outputs) and also the construction of the propositional networks.

Reliability and validity
No data regarding the reliability and validity of the technique are available. From previous experience, it is evident that the reliability of the technique may be questionable. Certainly, different analysts may identify different knowledge objects for the same scenario (Inter-analyst reliability). Also, the same analyst may identify different knowledge objects for the same scenario on different occasions (Inter-analyst reliability).

Tools needed
A propositional network analysis can be conducted using pen and paper. However, it is recommended that during the CDM procedure, an audio recording device is used. When constructing the propositional network diagrams it is recommended that Microsoft Visio is used.

Summary flowchart

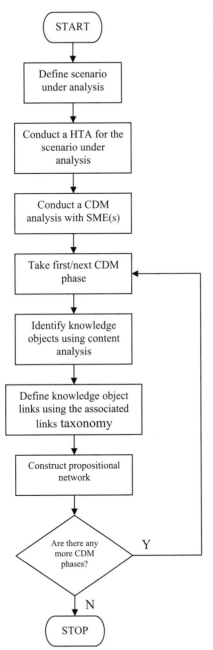

Summary

This chapter has presented a review of the EAST methodology and its component methods. The following conclusions can be drawn from the EAST methods review.

The EAST methodology is an exhaustive technique. A number of different analyses are conducted and various perspectives on the scenario(s) under analysis are offered. In its present form, the EAST methodology offers the following analyses of a particular C4i scenario:

- A step-by-step (goals, sub-goals, operations and plans) description of the activity in question.
- A definition of roles within the scenario.
- An analysis of the agent network structure involved (for example network type and density).
- A rating of co-ordination between agents for each team-based task step and an overall co-ordination rating.
- An analysis of the current technology used during communications between agents and also recommendations for novel communications technology.
- A description of the task in terms of the flow of information, communications between agents, the activity conducted by each agent involved and a timeline of activity.
- An analysis of agent centrality, sociometric status and betweenness within the network involved in the scenario.
- A definition of the key agents involved in the scenario.
- A cognitive task analysis of operator decision making during the scenario.

- A definition of the knowledge objects (information, artefacts etc.) required and the knowledge objects used during the scenario.
- A definition of shared knowledge or shared situation awareness during the scenario.

The EAST methodology is particularly suited to the analysis of team-based or collaborative activity, such as that seen in C4i environments. The method was originally developed for this purpose and applications so far have proved extremely successful, highlighting its suitability for such applications.

The EAST methodology is relatively simple to use. The method requires an initial understanding of HF and experience in the application of HF methods. However, from analyst reports it can be concluded that the EAST methodology is relatively simple to apply. The method's ease of use is heightened when compared to the exhaustive output that is generated from an EAST analysis.

The EAST methodology is generic and can be applied in any domain in which collaborative activity takes place. Despite the number of component methods involved in the EAST procedure, the method is generic and can be applied in any domain that uses collaborative activity.

Chapter 4

Case Study at HMS Dryad

With contributions from Roy Dymott, Rob Houghton, Geoff Hoyle, Mark Linsell, Richard McMaster, Paul S. Salmon, Rebecca Stewart, Guy H. Walker and Mark Young

Overview

This chapter describes a study conducted to analyse aspects of command and control in the Naval domain. Researchers were given access to a Royal Navy training establishment (the Cook simulation facility at HMS Dryad). Observations were made during Command Team Training (CTT), which involved training the Command Team of a warship in the skills that would be necessary for them to defend their ship in a multi-threat environment.

The aim of this study was to assess the communication and command on board the Type-23 frigate. The EAST methodology was used to explore a communication system that involved 16 team members where effective communication, decision making and coordination are essential to task success, for this highly distributed communication network.

Three scenarios (air threat, subsurface threat and surface threat) were observed and analysed. The scenarios were different, and due to the complexity of the task, observation focussed on individual crew members rather than the team as a whole. However an overall idea of communications could be ascertained from these individual observations.

Data are presented in the form of Social Network Analysis, Coordination Demand Analysis and Propositional Network Analysis. The propositional networks provide a way of exploring Shared Situation Awareness. Knowledge objects are identified for the mission and its phases. This gives an indication of where there is sharing of information and where knowledge is extended as the phases of the mission progress. The sharing of information could be as a result of direct communications between agents or through the use of the radio networks.

The individual scenarios show that the social networks are not particularly well distributed. However when the scenarios are amalgamated into one scenario as may happen in a real life threat it becomes a much denser network with better levels of participation. The network then becomes a split communications network with the AAWO-PWO and PWO being the central nodes.

Overall the analyses indicate that the Type-23 crew use a distributed network. Each crew member is connected (communication links) to other crew members and hence there are several channels with which communication can travel or information be shared. The results of this application of the methodology are intended to form a part of wider data collection and analysis with a view to developing a generic model of C4.

Table 4.1   Glossary of abbreviations

AAWO-PWO*	Anti Air Warfare Officer-PWO
APS	Air Picture Supervisor
ASonar	Active Sonar Team
ASWD	Active Systems Weapons Director
ASWPS	Anti Submarine Warfare Picture Supervisor
CO	Commanding Officer
CY	Communications Yeoman
EWD	Electronic Warfare Director
FC	Fighter Controllers
GD	Gun Director
HC	Helicopter Controller
MD	Missile Director
OOW	Officer of the Watch
PC	Picture Compilers
PSD	Passive Systems Director
PSonar	Passive Sonar Team
PWO	Principal Warfare officer
SPS	Surface Picture Supervisor

*The Type 23 Frigate does not have a dedicated AAWO, but rather it has a PWO with some restricted AAWO duties. For the sake of brevity, the AAWO-PWO is referred to as the AAWO throughout this chapter. In addition, during the observations at HMS Dryad, the AAWO-PWO was undertaking duties more commonly associated with conventional AAWO tasks.

Introduction

The Royal Navy has 16 Type-23 frigates. These were originally designed for the principle task of anti-submarine warfare but they have evolved into multi-purpose ships. In addition to their war fighting roles they are also used for embargo operations, disaster relief and surveillance operations.

The Royal Navy allowed a team of researchers access to one of their training establishments – the Maritime Warfare School – HMS Dryad in Southwick, Hampshire. Observations were made during Command Team Training (CTT). This programme involved training the Command Team of a warship in the skills that would be necessary for them to defend their ship in a multi-threat environment. These skills would enable the team to 'assimilate, interpret and respond correctly to the information received from external sources while reporting, directing and managing their and other units in the joint conduct of maritime operations' (Hoyle, 2001, p.3).

The training programme was conducted in a representative Type-23 'Operations Room Simulator' (ORS), as illustrated by Figure 4.1. The simulator room is slightly enlarged to allow for staff observation but otherwise was to scale (Hoyle, 2001). In addition to this simulator room there was a room that included a team of personnel who helped make any threats seem more realistic, that is, they portrayed the activity of other ships and aircraft as well as personnel from other parts of the ship.

Figure 4.1 **Illustration of workstations onboard a Type-23 frigate**

122   *Modelling Command and Control*

The scenarios

Three scenarios (air threat, subsurface threat and surface threat) were observed. The AAWO was the main agent observed in the air threat scenario and the PWO was the main agent observed for the subsurface and surface threat scenarios. Other agents although not directly observed were heard and seen interacting with either the AAWO or the PWO. The seating layout for the operations room can be seen in Figure 4.2. The AAWO would normally be plugged into the network at point*.

Command structure

Details of the agents involved in each scenario are presented in Table 4.2. Within the scenarios being studied there are four main agents. The officer of the watch (OOW) is an officer on the ship's bridge who maintains the visual lookout and controls the ship. The OOW can overrule the manoeuvring orders from the operations room should he consider them to be dangerous. The PWO is responsible for the tactical handling of the ship and the integrated use of its weapons systems and sensors. The PWO will take a tactical command role in multi threat missions. The AAWO is responsible for plan of defence in response to an air attack. The Captain will oversee the operations room. In addition to personnel the ship has a computer based command system that can communicate and control weapons and sensor systems, which allow information to be passed independently of the command system itself.

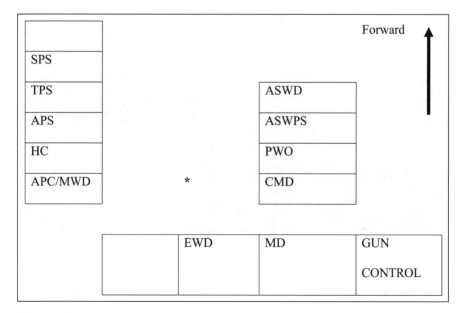

Figure 4.2   Seating lay out of Type-23 frigate operations room

Case Study at HMS Dryad 123

Table 4.2   The main agents involved in the mission

Agent Title	Details
Captain	Gives directions and instructions to PWO and AAWO
PWO	Principle Warfare Officer
AAWO	Anti Air Warfare Officer
EWD	Electronic Warfare Director – concerned with target and jammer status
MD	Missile Director – concerned with weapon and ammunition status, target engageability, firing and attack intentions
HC	Reviews helicopter status and is informed of attack intentions
APS/ Air	Air Picture Supervisor
Off Ship	Other ships, aircraft etc.
Duty Staff	Deals with external reports, platform and engagement status and sit reps.
Officer Watch (OOW)	Maintains the visual lookout and controls the ship
ASWD	Anti Submarine Warfare Director
ASWPS	Anti Submarine Warfare Picture Supervisor
SPS/ Surface	Surface Picture Supervisor
Harpoon	Firing and target data, firing authorisation
CY	Communication Yeoman
AcPS	Action Picture Supervisor

Communications

The majority of communication onboard the Type-23 frigate is conducted over radio circuits. The PWO and AAWO have access to 20 external circuits, of which there are (currently) four major nets. Internally, there is one main 'open line' with up to 40 point to point interphones. The subsurface and surface battles are usually fought over one net while the air battle is usually fought over another. The command open line is used by the captain, the AAWO, PWO, OOW, ASW Director, Duty Staff, EW Director, Missile and Gun Director, Communications Yeoman, and, on a part time basis, the helicopter controller.

Personnel are situated in quite close proximity to each other. The ASW, ASWD, PWO and captain are seated in a row as depicted in Figure 4.2 with the AAWO standing behind. This is advantageous as some of their observed communications were face-to-face, verbally or by way of written notes and pointing to the radar screens. Communications between the PWO and the remainder of the crew tended to be via the headsets. The AAWO was able to move freely around the operations room and again although communicated mainly using the headsets did verbally communicate face-to-face with several other crew members. The duty staff also tended to move freely around the operations room communicating via headset and face-to-face. Although only the PWO and AAWO were specifically observed it was noted that there was a lot of face-to-face communication between crew members seated adjacently.

Observations

The approach taken

Each member of the mission crew was seated at a console. An observer was able to sit directly behind a crew member and to plug into their console, allowing them to hear all radio exchanges. The AAWO was observed during the air threat and the PWO was observed during the subsurface and surface threats. All communication was noted. This included verbal exchanges not communicated via radio, hand gestures, and written communication (on paper). Once the observations were complete, the data collected were analysed using the Event Analysis of Systemic Teamwork (EAST) (Baber and Stanton, 2004) methodology. The scope of the analyses in this chapter is the examination of task and team interactions relevant to C4. The analysis does not focus in detail on the specific contextual make up of each mission.

Hierarchical Task Analysis

Hierarchical task analyses (HTA) (Annett and Stanton, 2000) of the Air threat operations scenario, the subsurface threat scenario and the surface threat were constructed using the observational data. Complete HTAs of the three scenarios are not presented in this chapter due to space constraints but are summarised by the following Task Model.

Task model

All three scenarios fit into the same task model. In order to manage this scenario, a number of goals need to be addressed: plan resources and strategy, control external resources, posture platform for attack, identify and classify targets, assess threat and allocate targets, engage targets and re-allocate assets and weapons. The task model shown in Figure 4.3 shows the relationship between these goals.

Before the mission, the planning of resources and strategy are undertaken. During the mission the central tasks are performed concurrently by different parts of the team. New targets are identified and classified by the picture compliers. The targets are then assessed and prioritised by the AWO and PWO, who allocate assets and weapon systems to the high priority threats. The targets are then engaged by the assets and weapon systems as appropriate, and the degree of success is assessed.

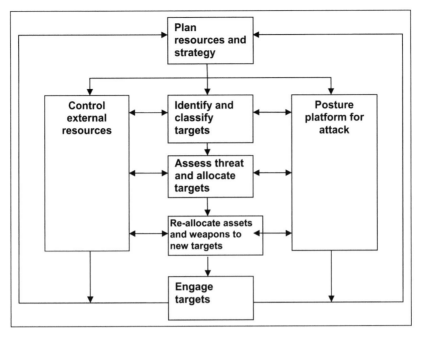

Figure 4.3 Combined task model for the air, subsurface and surface threat

Successfully damaging or deterring a target frees up the asset or weapon system for new allocation. Missed targets may require reallocation. At the same time as all of this activity is being undertaken, the platform is being postured to optimise the engagement or the ability to evade enemy weapons. There is also a requirement to coordinate with other platforms and control other external resources (such as fighters and helicopters). Thus the whole operation demands considerable coordination of both internal (to the platform) and external resources and assets to manage a mission and deal with threats.

Coordination Demand Analysis (CDA)

Both task-work (that is, task-oriented skills) and teamwork skills (that is, behavioural, attitudinal, and cognitive responses needed to coordinate with fellow team members) are needed in order to effectively complete team tasks.

Air threat CDA
The results are shown in Table 4.3. Key statistics from the CDA analysis show that approximately 58% of the HTA tasks are defined as task work, the remaining 42% as being related to team-work. Overall, the mean total coordination score is 2.45 (out of a maximum score of 3). Mean co-ordinations for the seven phases in the HTA are shown in Table 4.4. The majority of these co-ordinations are medium to high showing that there is a medium to high level of teamwork occurring. An extract of the CDA analysis is presented in Table 4.9.

Table 4.3 Air threat scenario CDA results

Category	Result
Total task steps	71
Total taskwork	41 (58%)
Total teamwork	30 (42%)
Mean Total Co-ordination	2.45
Modal Total Co-ordination	2.57
Minimum Co-ordination	1.00
Maximum Co-ordination	3

Table 4.4 Air threat scenario CDA results in HTA stages

	Plan resources	Detect and identify area movement	Engage targets	Control external resources	Posture platform in response to attack	Conduct damage assessments	Deploy decoys as appropriate	Total Mean
Mean comms	3.0	3.0	2.0	3.0	2.7	0	2.0	2.2
Mean SA	1.0	3.0	2.3	3.0	2.7	0	3.0	2.1
Mean DM	2.7	3.0	2.2	2.0	3.0	0	3.0	2.3
Mean MA	2.3	2.0	1.8	2.7	3.0	0	2.0	2.0
Mean leadership	2.7	2.1	2.2	2.7	3.0	0	2.0	2.1
Mean adaptability	2.1	2.1	2.3	2.7	2.7	0	2.0	2.0
Mean assertiveness	2.3	2.9	2.5	3.0	3.0	0	2.0	2.2
Total mean	2.3	2.6	2.2	2.7	2.9	0	2.3	-

Subsurface CDA

The results of the CDA for the subsurface threat are shown in full in Table 4.5, and the relevant total coordination scores are annotated directly on the OSD diagrams. Key statistics from the CDA analysis show that approximately 61% of the HTA tasks are defined as task work, the remaining 39% as being related to team-work. Overall, the mean total coordination score is 2.46 (out of a maximum score of three). Mean co-ordinations for the seven phases in the HTA are shown in Table 4.6. The majority of these co-ordinations are medium to high showing that there is a medium to high level of teamwork occurring.

Table 4.5   PWO subsurface scenario CDA results

Category	Result
Total task steps	71
Total taskwork	43 (61%)
Total teamwork	28 (39%)
Mean Total Co-ordination	2.46
Modal Total Co-ordination	2.57
Minimum Co-ordination	1.00
Maximum Co-ordination	3.00

Table 4.6   Subsurface threat scenario CDA results in HTA stages

	Plan resources	Detect and identify area movement	Respond to targets	Control external resources	Posture platform in response to attack	Conduct damage assessments	Defend ship	Total Mean
Mean comms	3.0	3.0	2.2	3.0	2.7	0	2.0	2.3
Mean SA	1.0	3.0	2.2	3.0	2.7	0	3.0	2.1
Mean DM	2.7	3.0	2.4	2.0	3.0	0	3.0	2.3
Mean MA	2.3	2.1	1.8	2.7	3.0	0	2.0	2.0
Mean leadership	2.7	2.0	2.4	2.7	3.0	0	2.0	2.1
Mean adaptability	2.1	2.1	2.2	2.7	2.7	0	2.0	2.0
Mean assertiveness	2.0	2.9	2.4	3.0	3.0	0	2.0	2.2
Total mean	2.3	2.6	2.2	2.7	2.9	0	2.3	-

Surface CDA

The results of the surface threat CDA are shown in full in Table 4.7, and the relevant total coordination scores are annotated directly on the OSD diagrams. Key statistics from the CDA analysis show that approximately 53% of the HTA tasks are defined as task work, the remaining 47% as being related to team-work. Overall, the mean total coordination score is 2.46 (out of a maximum score of 3). Mean co-ordinations for the nine phases in the HTA are shown in Table 4.8. The majority of these co-ordinations are medium to high showing that there is a medium to high level of teamwork occurring.

CDA conclusions

According to the CDA all three scenarios have a similar level of task work (Air: 58%; Subsurface: 61%; Surface: 53%) and team work (Air: 42%; Subsurface: 39%; Surface: 47%). The mean total coordination was virtually the same across all three scenarios (2.45; 2.46; 2.46 out of a possible 3; the highest level of teamwork). This represents a high level of coordination. A score of at least 75% or 2.25 is high and thus is indicative of good coordination.

The tables showing the HTA stages give an indication of where teamwork skills are prevalent. All three scenarios show a medium to high level of team working over the whole mission. The surface threat shows a particularly high level of teamwork for the phase 'reallocate new targets'. Where the scores are 0 this shows the phases that are deemed to be task work.

Social network analysis

Social network analysis (SNA) is used to analyse and represent the relationships existing between teams of personnel or social groups. The agents in the network are defined and their inter-connections specified. Interconnections are specified in terms of direction and in terms of strength.

The observations resulted in 'hub-and-spoke' type social networks. This is what one expects from having observations centralised on a single agent (that agent would function as the hub of the resulting network). Using all the communications data collected as well as talking to a subject matter expert (SME) a social network depicting the scenario as a whole was constructed (Figure 4.4). This diagram shows a split structure with sharing of information.

From the separate analyses of the AAWO and PWO, Figure 4.4 shows that the C2 team comprise a split network (Dekker, 2002). Split networks are identified with situations where there is a good communication and good intelligence. This would be a well distributed and hence a denser network indicating higher rates of participation. As can be seen the PWO and AAWO form the main hubs of the network. The majority of crew members communicate with both of these agents, however there are some crew members who only communicate with one of them. Although this figure indicates that the AAWO and PWO have an equal amount of communications, from the observations the PWO communicated with more crew members more frequently than the AAWO.

Table 4.7   PWO surface scenario CDA results

Category	Result
Total task steps	68
Total taskwork	36 (53%)
Total teamwork	32 (47%)
Mean Total Co-ordination	2.46
Modal Total Co-ordination	2.57
Minimum Co-ordination	1.00
Maximum Co-ordination	3.00

Table 4.8   Surface threat scenario CDA results in HTA stages

	Plan resources and strategy	Detect and identify area movement	Engage targets	Control external resources	Determine defence type	Posture platform in response to attack	Reallocate new targets	Conduct damage assessments	Deploy decoys as appropriate	Total Mean
Mean comms	3.0	2.8	1.7	3.0	0	2.7	3.0	3.0	2.0	2.4
Mean SA	1.0	2.9	2.0	3.0	0	2.7	3.0	2.0	2.0	2.3
Mean DM	2.7	2.8	2.0	2.0	0	3.0	3.0	3.0	2.0	2.3
Mean MA	2.3	1.9	2.0	2.7	0	3.0	3.0	2.0	2.0	2.3
Mean leadership	2.7	2.2	2.0	2.7	0	3.0	3.0	2.0	3.0	2.3
Mean adaptability	2.1	2.3	2.0	2.7	0	2.7	3.0	2.0	3.0	2.2
Mean assertiveness	2.3	2.9	2.0	3.0	0	3.0	3.0	2.0	2.0	2.2
Total mean	2.3	2.5	2.0	2.7	0	2.1	3.0	2.3	2.3	-

Table 4.9   Extract of CDA analysis

Task Step	Agent	Activity	Task
1		**Plan resources and strategy**	
1.1		**Determine mission plans**	
1.1.1	AAWO/EWD	**Determine overt plan**	
1.1.2	AAWO/EWD	**Determine covert plan**	
1.1.3	AAWO/EWD	**Determine day plan**	
1.1.4	AAWO/EWD	**Determine night plan**	
1.1.5	AAWO/EWD	**Determine satellite passes**	X
1.1.6	AAWO/EWD	**Choose plan**	
1.2	AAWO	**Determine immediate priorities**	
1.3	AAWO	**Conduct general briefing and discussion**	
1.4	AAWO	**Conduct threat analysis**	
1.5	AAWO	**Balance threats to assets over given timeline**	X
1.6	AAWO	**Draw up plan**	
1.7	AAWO	**Order ind specialists to write up plan execution instructions**	X
1.8	AAWO	**Compile operational order to guide activity**	X
2		**Detect and identify area movement**	
2.1	Picture compilers	**Detect targets**	
2.2		**Identify/classify air objects**	
2.2.1	Picture compilers	**Classify as friendly**	
2.2.2	Picture compilers	**Classify as neutral**	
2.2.3	Picture compilers	**Classify as unknown**	

Team	Communication	Situational Awareness	Decision Making	Mission Analysis	Leadership	Adaptability	Assertiveness	Total Coordination
X	3	1	3	3	3	2	2	2.43
X	3	1	3	3	3	2	2	2.43
X	3	1	3	3	3	2	2	2.43
X	3	1	3	3	3	2	2	2.43
X	3	1	3	3	3	3	3	2.71
X	3	1	3	1	2	2	3	2.14
X	3	1	2	1	3	2	3	2.14
X	3	1	2	2	2	2	2	2.00
X	3	1	2	2	2	2	2	2.00
X	3	3	3	3	2	3	2	2.71
X	3	3	3	2	2	2	3	2.57
X	3	3	3	2	2	2	3	2.57
X	3	3	3	2	2	2	3	2.57

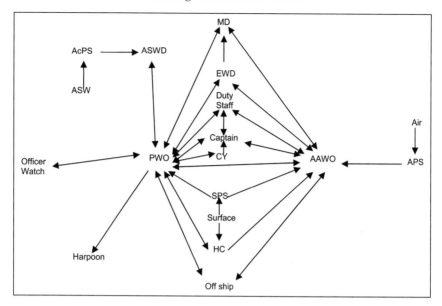

Figure 4.4   Social network encompassing all three scenarios

Operation sequence diagram

Operation sequence diagrams (OSDs) can be used to provide a pictorial representation of a task. Separate steps of a task can be highlighted along a time line. The OSD uses icons and labels that are standardised for this method and captures the flow of information among distributed actors. Three OSDs were constructed from the data that was gathered during the observation of the three scenarios and are shown in Figures 4.5 to 4.7. The circles represent a transmission.

Air threat

From the OSD for the Air threat scenario a table summarising the loading on each agent within the network can be produced. This is shown in Table 4.10. Only receive and transmit were identified during the observations. Decision making was occurring all the time but due to the fast paced environment it was impossible to note all these down.

From the operational loadings it can be seen that during communications the AAWO has the highest loading, with the EWD and captain having similar but much lower loadings. The AAWO was extremely busy transmitting information. This reiterates what was found in the social networks for the air threat task in that the AAWO was the main node during this scenario. The OSD of the air threat is shown in Figure 4.5.

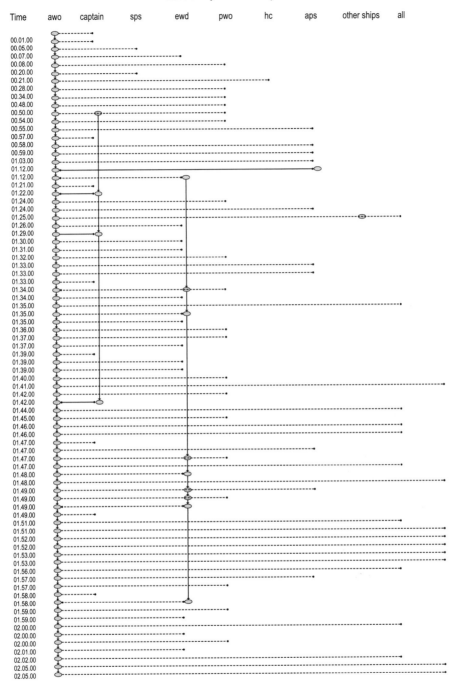

Figure 4.5 OSD air threat scenario

Subsurface threat

The OSD of the subsurface is shown in Figure 4.6. From the OSD for the subsurface threat scenario a table summarising the loading on each agent within the network can be produced. This is shown in Table 4.11. Similarly to the air threat scenario only receive and transmit were identified during the observations. Decision making was occurring all the time but due to the fast paced environment it was impossible to note all these down.

Table 4.10   Operational loading for the air threat scenario

Agent	Receive	Transmit
AAWO	7	88
PWO	9	
Captain	8	4
SPS	2	
EWD	6	10
HC	1	
APS	9	1
Other ships		1

Table 4.11   Operational loading for the subsurface threat scenario

Agent	Receive	Transmit
AAWO	3	6
PWO	61	105
Captain	15	17
ASW	8	3
ASWD	13	5
HC	12	10
Officer Watch	28	9
MD	1	1
EWD	3	7
APS		1
SPS/Surface	1	
Duty Staff	3	2
Other Ships	10	2

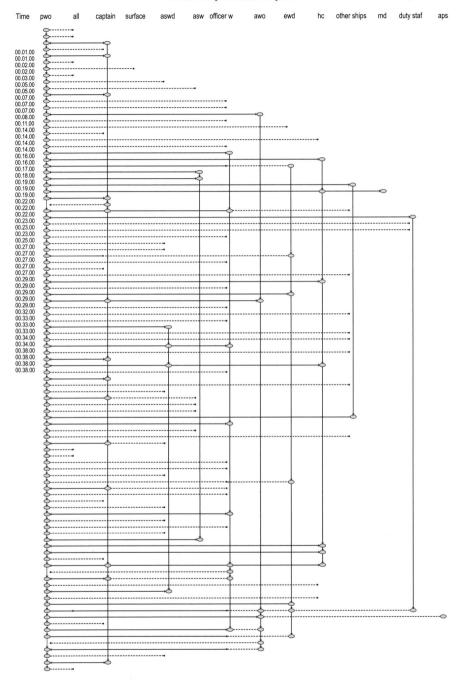

Figure 4.6 OSD subsurface threat scenario

From the operational loadings it can be seen that during communications the PWO has the highest loading, with the Captain and officer of the watch having similar but much lower loadings. The PWO was extremely busy transmitting information as well as receiving information. This reiterates what was found in the social networks for the subsurface threat task in that the PWO was the main node during this scenario.

Surface threat

From the OSD for the surface threat scenario a table summarising the loading on each agent within the network can be produced. This is shown in Table 4.12. Only receive and transmit were identified during the observations. Decision making was occurring all the time but due to the fast paced environment it was impossible to note all these down.

From the operational loadings it can be seen that, similarly to the subsurface task, during communications the PWO has the highest loading, with the ASW and officer of the watch having similar but much lower loadings. The PWO was extremely busy transmitting information as well as receiving information. This reiterates what was found in the social networks for the surface threat task in that the PWO was the main node during this scenario. The OSD of the Surface threat is shown in Figure 4.7.

OSD conclusions

The OSDs give a graphical representation of each of the scenarios with the circles representing transmissions in communication. Unfortunately OSDs do not always represent large, complex tasks as well as simpler ones. It was noticed that while observing the scenario a large amount of the communications were in the form of

Table 4.12 Operational loading for the surface threat scenario

Agent	Receive	Transmit
AAWO	1	
PWO	26	51
Captain	8	5
ASW	12	3
ASWD	2	
Harpoon	4	4
Officer Watch	15	4
EWD	4	7
Air	1	
Surface	1	
Duty Staff		3

questions and these have not been accounted for in the symbols. Decisions were made throughout the three scenarios, however they could not be pictorially represented on the OSDs either.

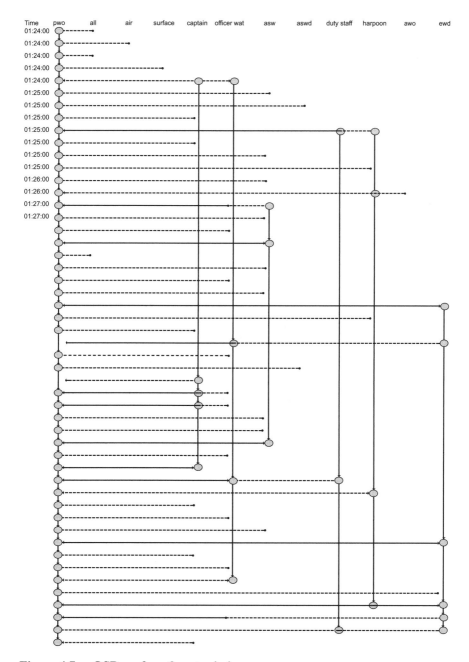

Figure 4.7 OSD surface threat mission

The operational loadings for the three scenarios reiterated what was found in the social network analysis. The AAWO was the predominant communicator for the air threat while the PWO was the predominant communicator during the surface and subsurface scenarios. However this may be due to the fact that the AAWO was observed during the air threat and the PWO during the subsurface and surface threat. However if the overall social network is taken into account the AAWO and PWO still remain the main hubs and therefore it would be fair to assume that with respect to the receiving and transmitting of information the OSDs are representative.

Critical decision method

The critical decision method (CDM) requires a structured interview to elicit responses against defined categories. A subject matter expert was interviewed and CDM analyses were conducted for the AAWO air threat scenario, PWO subsurface threat and PWO surface threat scenario. The probes defined by O'Hare et al. (2000) were used to evaluate the decision making during the scenario. The CDM probes used were described in chapter three.

The air threat scenario was divided into eight phases, 'planning resources and strategy', 'controlling external resources', 'identifying and classifying targets', 'assessing threats and allocating targets', ' engagement of targets', 'posture platform and response to an attack', 'reallocate assets to new targets', and 'assess damage to target'.

The subsurface threat was divided into six phases, 'planning resources and strategy', 'conduct threat identification', 'evade and engage targets', 'controlling external resources', 'posture platform and response to an attack', and 'assess damage to target'.

The surface threat was divided into six phases, 'planning resources and strategy', 'conduct threat identification', 'evade and engage targets', 'controlling external resources', 'posture platform and response to an attack', and 'assess damage to target'.

From the three CDM analyses, propositional networks were constructed in order to identify the knowledge objects (sources of information, agents and objects) involved in the scenarios. The knowledge objects identified represent the ideal collection of knowledge objects for the scenarios that were used in the CDM analyses. The propositional networks identified a total of 60 knowledge objects for all three scenarios.

Propositional networks

Propositional networks analysis
From the CDM and discussion it is possible to construct an initial propositional network to show the knowledge that is related to this incident. The propositional network consists of a set of nodes that represent sources of information, agents, and objects etc. that are linked through specific causal paths. From this network, it should be possible to identify required information and possible options relevant to this incident. The concept behind using a propositional network in this manner is that

it represents the 'ideal' collection of knowledge for an incident (and is probably best constructed post-hoc). As the incident unfolds, so participants will have access to more of this knowledge (either through communication with other agents or through recognising changes in the incident status). Consequently, within this propositional network, Situation Awareness can be represented as the change in weighting of links.

Propositional networks of the three scenarios are presented in the following pages. In all cases the overall propositional network is presented first followed by smaller figures representing the individual phases within each scenario. Although these are quite small, it is clear how the pattern of knowledge objects changes between the phases. Larger versions of these figures can be found in Stanton et al. (2006).

Figure 4.8 shows the propositional network for the air threat scenario. Figures 4.9 to 4.15 represent the propositional networks for the individual phases of the mission. These phases coincide with the phases in the task network. Figure 4.16 shows the shared awareness of the air threat mission (shaded objects).

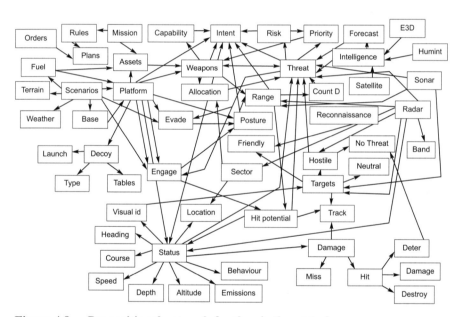

Figure 4.8   Propositional network for the air threat task

140 *Modelling Command and Control*

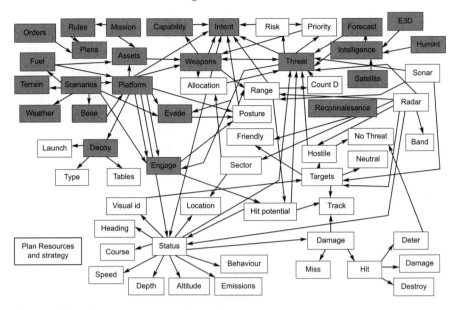

Figure 4.9   Plan resources and strategy

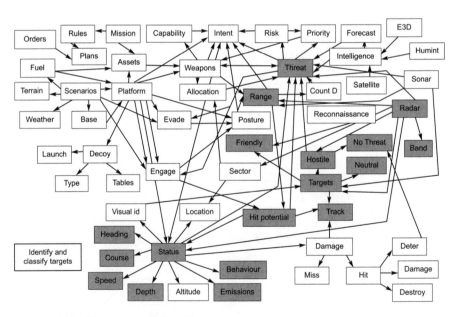

Figure 4.10   Identify and classify targets

Case Study at HMS Dryad 141

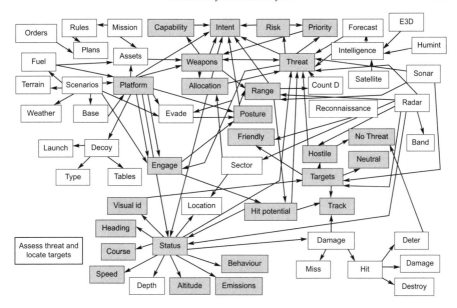

Figure 4.11 Assess threat and locate targets

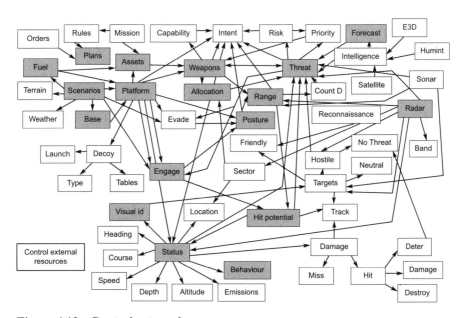

Figure 4.12 Control external resources

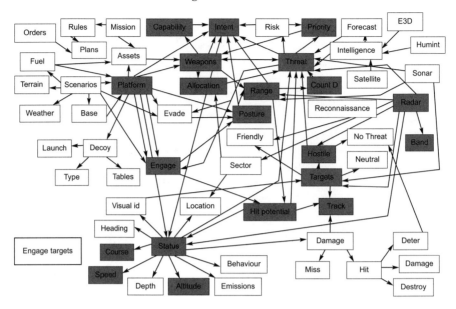

Figure 4.13 Engage targets

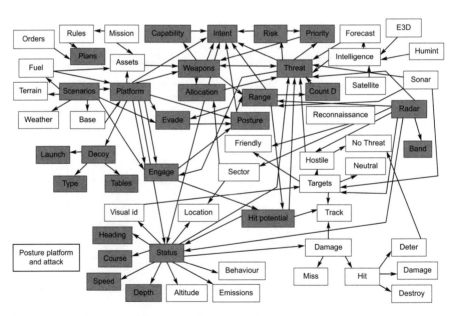

Figure 4.14 Posture platform and attack

Case Study at HMS Dryad 143

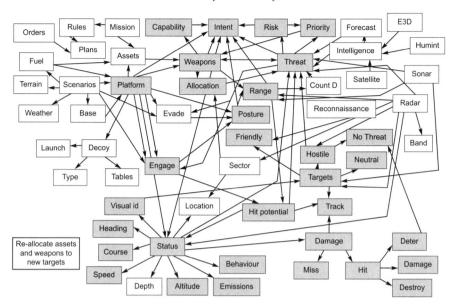

Figure 4.15 **Re-allocate assets and weapons to new targets**

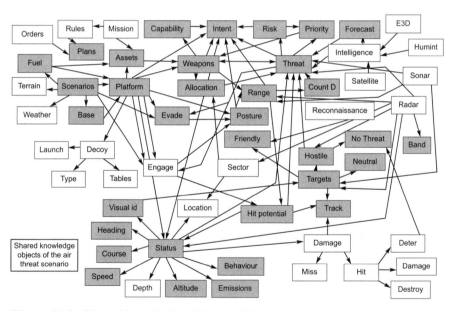

Figure 4.16 **Shared knowledge objects of the air threat scenario**

Similarly Figure 4.17 is the overall propositional network for the subsurface threat whilst Figures 4.18 to 4.24 show the steps of the task. Figure 4.25 shows the overlap of knowledge objects for the subsurface threat.

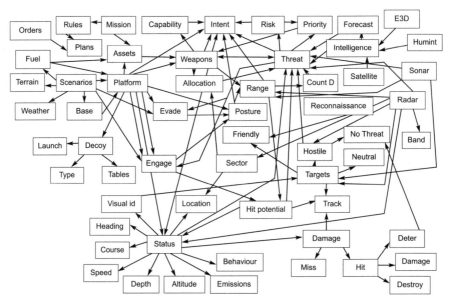

Figure 4.17 Propositional network for the sub-surface threat task

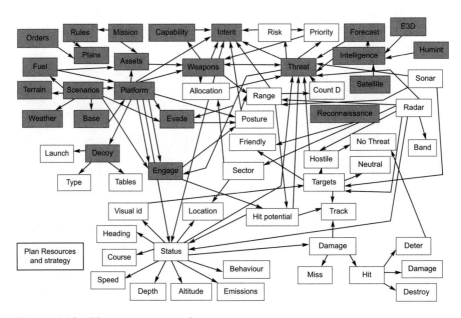

Figure 4.18 Plan resources and strategy

Case Study at HMS Dryad 145

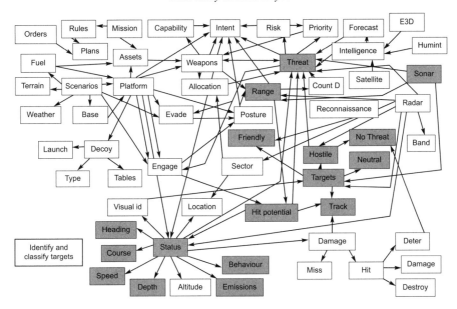

Figure 4.19 Identify and classify targets

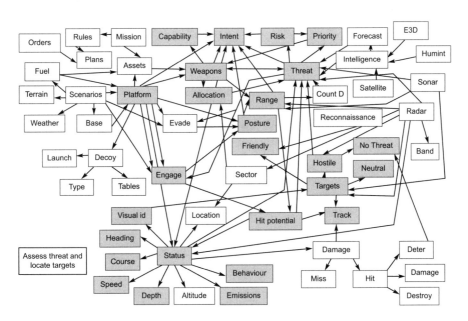

Figure 4.20 Assess threat and locate targets

146   *Modelling Command and Control*

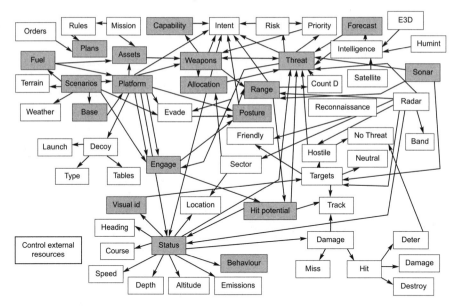

Figure 4.21   Control external resources

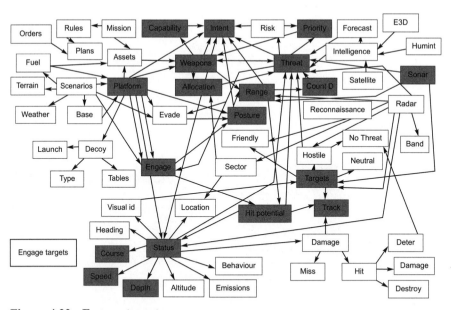

Figure 4.22   Engage targets

Case Study at HMS Dryad   147

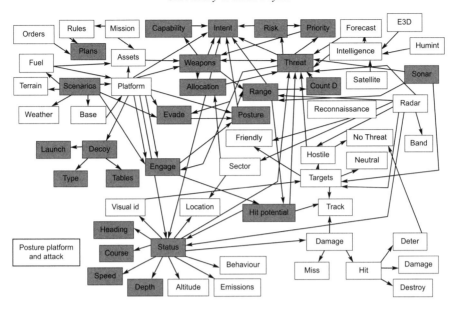

Figure 4.23   Posture platform and attack

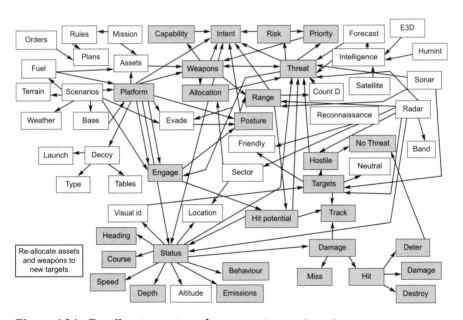

Figure 4.24   Re-allocate assets and weapons to new targets

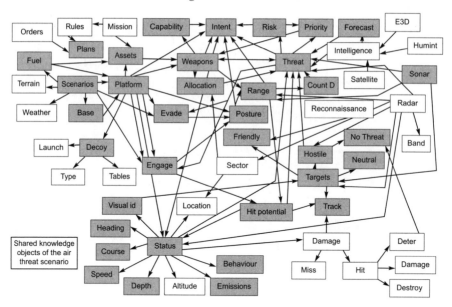

Figure 4.25 Shared knowledge objects of the air threat scenario

Finally, Figure 4.26 shows the overall propositional network of the surface threat and Figures 4.27 to 4.33 show the breakdown of phases. Figure 4.34 shows the overlap of knowledge objects for the surface threat scenario.

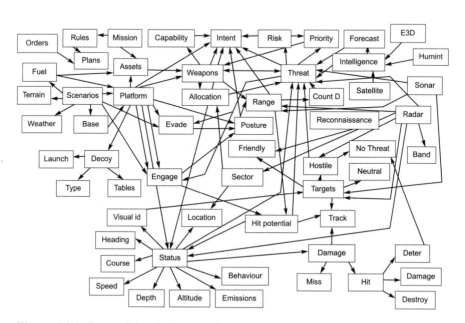

Figure 4.26 Propositional network for the surface threat task

Case Study at HMS Dryad 149

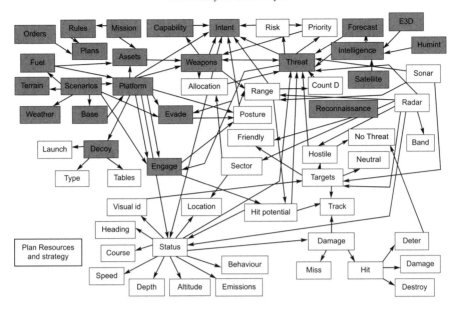

Figure 4.27 Plan resources and strategy

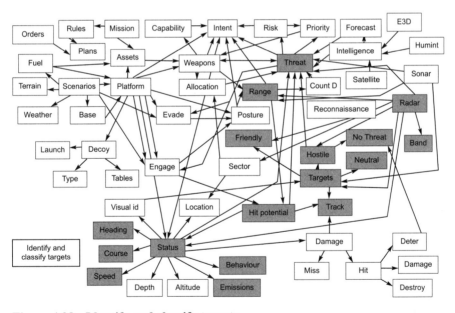

Figure 4.28 Identify and classify targets

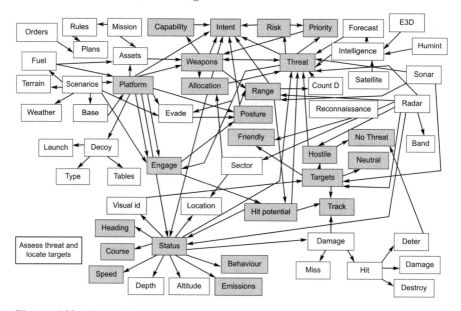

Figure 4.29 Assess threat and locate targets

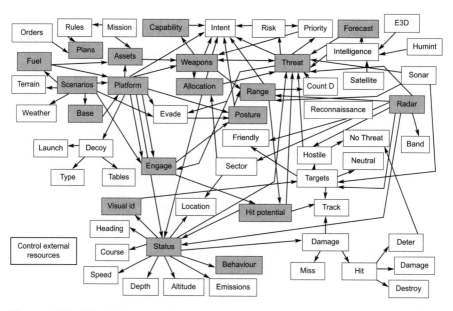

Figure 4.30 Control external resources

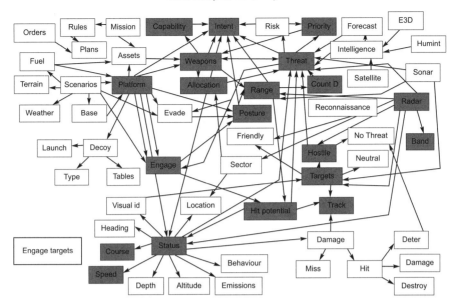

Figure 4.31 Engage targets

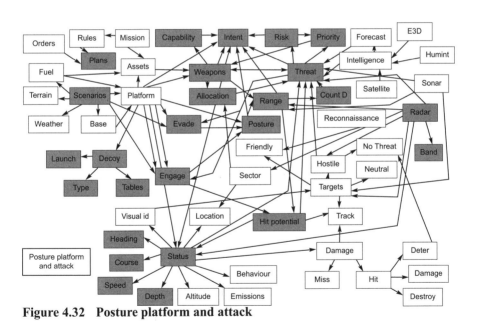

Figure 4.32 Posture platform and attack

152   *Modelling Command and Control*

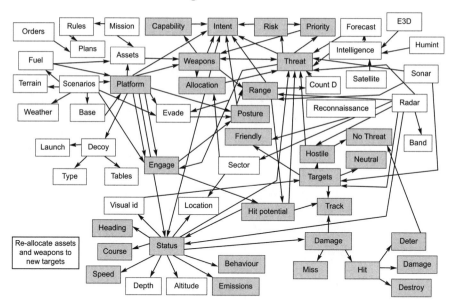

Figure 4.33   **Re-allocate assets and weapons to new targets**

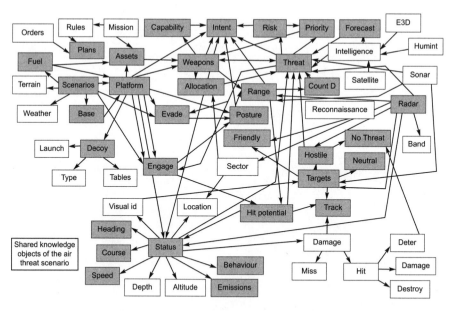

Figure 4.34   **Shared knowledge objects of the air threat scenario**

Propositional network conclusions

The Propositional networks for the three 'threat' tasks (that is air, surface and subsurface) were constructed by identifying the knowledge objects in the CDM tables. A propositional network was constructed separately for each of the tasks. Content validation of the propositional networks was performed by reference to the HTA (for cross-validation) and a subject matter expert (for content validation). As the propositional networks deal with the representation of knowledge, rather than specific technologies, there is little difference between the three analyses. The main difference is between the subsurface and air networks, as the subsurface network refers to 'sonar' and 'depth' objects, whereas the air network refers to 'radar' and 'altitude' objects. To simplify the analyses, a single network has been presented, comprising both types of knowledge object, but only the knowledge objects relevant to each scenario are highlighted.

A further simplification of the propositional network has been the removal of the labels in order to make the diagrams readable. This is simply to make the propositional network readable and fit onto a single page. All three threat tasks have the same seven phases of operation associated with them, namely: planning resources and strategy, identify and classify targets, assess threat and allocate targets, engage targets, reassess targets and reallocate weapons to targets, control external resources, and posture platform for attack. The knowledge objects pertinent to each phase have been highlighted in each of the propositional networks.

From this analysis, it is possible to identify the key knowledge objects that have salience to each phase of operation. For the purpose of this report, salience is defined as those knowledge objects that serve as a central hub to other knowledge objects (that is, have five or more links to other knowledge objects). This criterion produces a list of 12 knowledge objects from a pool of 64 (approximately one fifth of the total number of knowledge objects). The objects are: intent, weapons, scenarios, threat, range, engage, radar, targets, status, intelligence, platform and hit potential. The purpose of this analysis is to identify knowledge objects that play a central role in the threat tasks. Each of these core knowledge objects is represented in a generic table against each stage for the purpose of highlighting its role.

As Table 4.13 shows, different core knowledge objects are salient at different points in the operation (for example, intent is relevant at the plan, allocate, engage, reassess and posture phases, whereas intelligence is relevant only in the plan phase). The passing of knowledge objects from one phase to another involves some manipulation of the object before it is passed and then a means of communicating the nature of the object (for example the priority of targets are assessed before weapon systems are allocated to them and then they may be engaged), either implicitly or explicitly.

Table 4.13 Analysis of core knowledge objects against the seven phases of operation

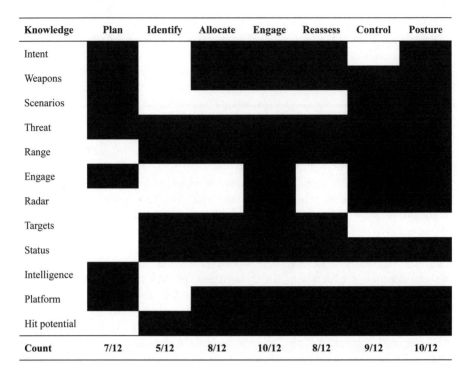

Knowledge	Plan	Identify	Allocate	Engage	Reassess	Control	Posture
Intent	■		■	■	■		■
Weapons	■			■	■	■	■
Scenarios	■			■		■	■
Threat	■	■	■	■	■	■	■
Range	■		■	■		■	■
Engage	■		■	■	■		■
Radar		■	■	■	■	■	■
Targets		■	■	■	■		■
Status		■	■	■	■	■	■
Intelligence	■					■	■
Platform			■	■	■	■	■
Hit potential		■	■	■	■	■	■
Count	**7/12**	**5/12**	**8/12**	**10/12**	**8/12**	**9/12**	**10/12**

Conclusions

In this chapter the various representations are shown and their construction illustrated. Overall it can be seen that the scenarios observed were very intense and quickly changing with an enormous amount of communications occurring over a relatively short period of time.

One issue was how representative the data was on account of observer expertise. Some information may not have been recorded due to a lack of proficiency in listening to multiple channels of information. It should also be remembered that the observed scenarios were part of a training course and, although realistic, there may have been more communications between certain crew members due to this training. A subject matter expert (SME) was available to clarify any queries and a large amount of data were collected for each of the three scenarios.

The HMS Dryad scenarios were a highly complex interaction of crew and communication channels. It is an extremely intense environment and over a relatively short period of time (approximately two hours) an enormous amount of communications occurred with information being transferred in seconds.

The methods indicate that there is a lot of teamwork occurring in each scenario although there is a clear hierarchy. Although the scenarios were analysed independently they could all be occurring at the same time in a real event. It is not surprising therefore that all three scenarios showed similar levels of task and team work. Where team work was necessary the scenarios showed a high level of coordination. According to the CDAs there was slightly more team work occurring in each of the three scenarios. However the values for teamwork were not considerably higher than for task work.

Spoke diagrams illustrate the social networks for each scenario. These diagrams show that the social networks are not particularly well distributed, thus communication between all agents can be seen as poor. This, however, is due to the fact that only one crew member's communications were observed. However if we look at the mission as a whole encompassing all three scenarios and with the advice of a subject matter expert it can be seen that it becomes a much more split network as opposed to the spoke networks with better levels of participation. The PWO and AAWO still remain the central nodes of the operations room. Information is shared between the crew members, however the majority of this information seems to be shared via the PWO and AAWO.

Shared awareness can be seen from the propositional networks. Many of the knowledge objects are shared within the three individual scenarios as well as across the whole mission. It is important to remember that the three scenarios observed will often happen at the same time and will not be separated into three clear areas. Thus the sharing of knowledge objects across scenarios will be essential for effective operations.

Overall the analyses indicate that the Type-23 crew use a distributed network. Each crew member is connected (communication links) to other crew members and hence there are several channels with which communication can travel or information be shared. The results of this application of the methodology are intended to form a part of a wider data collection and analysis with a view to developing a generic model of C4.

Chapter 5

Case Study in RAF Boeing E3D Sentry

With contributions from Alison Kay, Mel Lowe, Paul S. Salmon, Rebecca Stewart, Kerry Tatlock and Linda Wells

Introduction

The observations took place on board a Royal Air Force (RAF) Boeing E3D Sentry AWACS (Airborne Warning and Control System) aircraft. The RAF ran a two-week training course for Combined Qualified Weapons Instructors (CQWI). A simulated war exercise was carried out each day involving three key teams of personnel: ground based support, the E3D (AWACS) team and the fighter pilots. Ground based support includes all personnel assigned to the mission who are not flying in the simulated war exercise. The term 'fighter pilots' refers to both fighter and bomber pilots who took part in the exercise. These pilots flew between 20 and 40 aircraft for each mission.

The role of the E3D team was to provide support for both ground and fighter personnel, and involved providing a global picture of the war from the sky as it developed. This information was relayed to ground support staff and to individual fighter pilots. All personnel were kept up to date with fighters' positions in relation to one another and of any fatalities.

There are 18 crew members who operate the aircraft and make surveillance and support for ground and fighters possible (seating plan shown in Figure 5.1). The 18 E3D crew can be summarised as follows:

- Flight crew: Pilot, 1st Officer, Navigator and Flight Engineer.
- Tactical Director.
- Surveillance Team: Surveillance Controller, two Surveillance Operators and the Electronic Surveillance Measures Operator – these crew members are involved with identifying aircraft and tracking their movement.
- Weapons Team: Fighter Allocator and three Weapons controllers (Strike, Check-in and OCA) – these crew members are responsible for relaying information to the fighters on the whereabouts of the enemy, whether any aircraft have been shot down. Check-in is responsible for authenticating the fighters. The OCA is responsible for liasing with the fighters and informing them of the enemy's whereabouts. The strike directs the use of weapons.
- Link Manager.
- Technicians: Communications Operator, Communications Technician, Display Technician and Radar Technician.

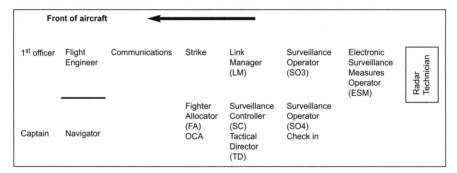

Figure 5.1   Typical seating plan onboard an RAF E3D

Command structure for the E3D mission crew

The E3D crew are directed by a Tactical Director (TD). The TD will generally work through the Surveillance Controller (SC) and the Fighter Allocator (FA) although communications do occur between the TD and other crew members. The FA will have the weapons controllers reporting to them. The weapons controllers will consist of check-in, strike and OCA. The SC will have the two surveillance operators (SO3/SO4) and the link manager (LM) reporting to them. Figure 5.2 shows the structure of the mission crew. The TD should not be over-loaded: they are there to supervise the mission, to assess the situation and to stop accidents. They will monitor work levels and will be thinking ahead to assess possible problems and their solutions. The SC and the FA will make decisions but will usually refer to the TD.

Communications

The majority of communications onboard the AWACS are verbal using either the three internal networks or four external radios. All crew are able to talk to anyone on board using the network including the flight crew. The networks are used for the mission crew and flight deck to interact, by the weapons team and by the surveillance team. Some information is passed verbally face-to-face although this is usually only between those crew members who are sitting next to each other or those who can turn round to face each other. Hand gestures, particularly 'thumbs up' and 'pointing' are also frequently used as are nodding of heads. Figure 5.2 indicates the hierarchical passage of information. The majority of the surveillance team's communications will remain on board the E3D whereas the weapons team will communicate off the aircraft, particularly with the fighters and bombers.

In addition to verbal communications, information is passed via the radar screens. This can be in the form of text messages that can be sent and received by all crew members. Arrows are also used on the screens to highlight aircraft. Probably the most important way of sharing information is when one crew member identifies an aircraft and amends the system, which is then interpreted by other crew members.

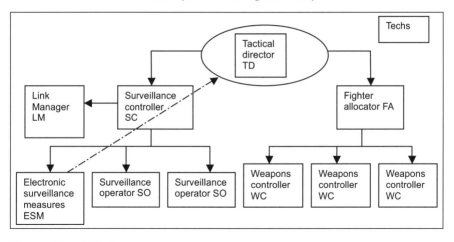

Figure 5.2 **Mission crew structure**

The scenario

A different mission, with different objectives and rules of engagement, was run for each day of the training period. A mass briefing took place each day providing all personnel with information specific to that day: where and when the exercise would be taking place, weather conditions, available resources, who would be taking part and what they would be expected to do. Personnel broke off into their respective teams and planned the mission. Further briefs were given (as and when dictated by the training co-ordinators) in order to develop the scenario.

The E3D was always the first aircraft in the sky and the last to return to base. The aircraft would circle its designated area and the crew have the systems up and running, ready to support all personnel when the war began. The point at which the aircraft was in position and the crew were ready to support the mission was termed 'on station' and it is from this point that the observations began for each mission. Observations continued throughout the war and were terminated when the crew were 'off station' (that is when the war was over) and ready to return to base. Debriefs (both mass and team debriefs) took place after each mission in order to digest the events of that day.

Observations

The approach taken

Observers had the opportunity to attend briefings and debriefings and were given access to relevant paperwork for each mission. Two observers were permitted to be onboard for each mission. Each observer monitored the communications for one team member for the duration of the mission.

Each member of the mission crew was seated at a console (as shown in Figure 5.3) and had access to three internal networks and four external radio channels, which they could listen to simultaneously. Observers were able to stand directly behind mission crew (viewing their display screen) and to plug into their console, allowing them to hear all radio exchanges. All forms of communication were taken note of. This included verbal exchanges not communicated via radio, text messaging (from console to console), hand gestures, written communication (on paper) and graphical prompts (arrows sent from console to console and displayed on screen).

Task model

In order to manage this scenario the crew need to deal with the following goals: assume station, ensure operational procedures are followed, manage self defence, manage operational/comms security, manage air refuelling, manage emergency procedures, manage comms, control surveillance, operate electronic surveillance manager (ESM) coordinate crew and manage airborne battle. The task model shown in Figure 5.4 shows the relationship between these goals.

Coordination demand analysis

Both task-work (that is task-oriented skills) and teamwork skills (that is behavioural, attitudinal, and cognitive responses needed to coordinate with fellow team members) are needed in order to effectively complete team tasks. The CDA procedure allows the identification of the operational skills needed within team tasks, but also the teamwork skills needed for smooth coordination among team members. Teamwork skills were extracted from the HTA and rated against the CDA taxonomy of: communication; situational awareness; decision making; mission analysis; leadership; adaptability; and assertiveness. Each stage of the CDA is scored from 1 to 3 where 1 is low coordination and 3 is high coordination. A score of at least 75% or 2.25 is fairly high and thus is indicative of good co-ordination.

From these individual scores a 'total coordination' figure can be derived. The results are shown in full in Table 5.1. Key statistics from the CDA analysis show that approximately 20% of the HTA tasks are defined as task work, the remaining 80% as being related to team-work. Overall, the mean total coordination score is 2.23 (out of a maximum score of 3). Mean co-ordinations were also calculated for the 11 main stages of the HTA and are shown in Table 5.2. The majority of these co-ordinations are medium to high showing that there is a medium to high level of teamwork occurring. An extract of the CDA analysis is presented in Table 5.3.

Social network analysis (SNA)

Social network analysis (SNA) is used to analyse and represent the relationships existing between teams of personnel or social groups. The agents in the network are defined and their inter-connections specified. Interconnections are specified in terms of direction and in terms of strength. The E3D mission has been broken down into nine agents. The agents involved are presented in Table 5.4.

Case Study in RAF Boeing E3D Sentry

Figure 5.3 Illustration of workstations on board the E3D

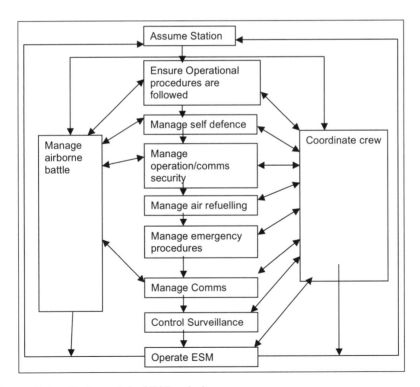

Figure 5.4 Task model of E3D mission

Table 5.1   E3D mission scenario CDA results

Category	Result
Total task steps	292
Total taskwork	56 (20%)
Total teamwork	236 (80%)
Mean Total Co-ordination	2.23
Modal Total Co-ordination	2.86
Minimum Co-ordination	1.00
Maximum Co-ordination	3.00

Table 5.2   E3D mission CDA results in HTA stages

Category	Assume Station	Manage Airborne Battle	Co-ordinate crew	Manage communications	Ensure operational procedures followed	Manage self defence	Manage comms/ operation security	Manage air refuelling	Manage emergency procedures	Control surveillance	Operate ESM equipment
Mean Comms	2.5	2.2	2.0	1.9	2.2	2.6	1.8	2.0	1.9	2.0	1.4
Mean SA	2.8	2.9	2.8	2.3	3.0	3.0	2.5	3.0	2.9	2.9	2.0
Mean DM	2.2	2.4	2.5	2.0	2.7	2.8	1.8	2.0	2.6	2.0	1.2
Mean MA	2.2	2.5	2.5	1.8	2.7	3.0	1.3	1.0	2.6	1.9	1.1
Mean Leadership	2.2	2.4	2.7	1.5	2.8	3.0	2.5	2.0	3.0	2.2	1.0
Mean Adaptability	2.2	2.9	2.7	2.1	3.0	3.0	3.0	3.0	3.0	2.8	1.8
Mean Assertiveness	2.5	2.9	2.7	2.3	3.0	3.0	3.0	3.0	3.0	2.8	1.9

From the list of agents that have been identified for this particular system, a matrix of association can be constructed (Table 5.5). The list of agents has been condensed from 18 roles into 9. This is a result of not every mission crew member being observed and an attempt to establish a representative evaluation of the overall team and task objectives for C4. Assumptions made in order to do this are as follows:

- All Weapons Controllers will be represented by 'weapons controller' regardless of their mission task (Strike, Check-in or OCA).
- All Surveillance Operators will be represented by 'surveillance operator' regardless of their mission task (surveillance operators, electronic surveillance operator).
- All technicians will be represented by 'technician' regardless of their mission task (communications operator, communications technician, display technician, radar technician).
- All external parties will be represented by 'off aircraft'.
- All members of the flight crew will be represented by 'cockpit' regardless of their mission task (pilot, 1st officer, navigator, flight engineer).
- Off aircraft refers to communications with air traffic control and predominantly to fighter and bomber aircraft particularly for the weapons controllers.

This matrix shows whether or not an agent can be associated with any other agent, specifically through communications where 0 is no communication and 1 indicates that the agents communicated.

In addition to the matrix a social network diagram can be created. The social network diagram illustrates the proposed association between agents on board the E3D. The social network diagram for the E3D mission is presented in Figure 5.5.

The number of crew members observed was limited therefore a generic social network diagram was constructed using the standard operating procedures (SOPs).

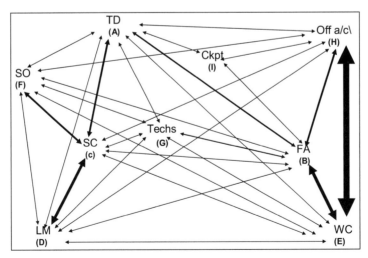

Figure 5.5 Social network diagram constructed from the standard operating procedures

Table 5.3 Extract of CDA analysis where 1 = low, 2 = medium and 3 = high

Task Step	Agent	Activity
1.0	TD	**Assume station**
1.1	TD	**Follow checklists and procedures**
1.2	TD	**Execute safe joining procedures**
1.3	TD	**Communicate with TACON**
1.4	TD	**Ensure radar modes are optimised**
1.5	TD	**Confirm crew are ready to go on station**
1.6	TD	**Determine on station time**
2.0	TD	**Manage Airborne Battle**
2.1	TD	**Coordinate support of assets**
2.2	TD	**Minimise risk to friendly forces**
2.3	TD	**Direct the air battle**
2.3.1	TD	**Implement airspace control plan**
2.3.2	TD	**Implement airspace regulation procedures**
2.3.3	TD	**Implement published identification procedures**
2.3.4	TD	**Implement published Rules of Engagement**
2.3.5	TD	**Monitor tactical situation**
2.3.6	TD	**Coordinate allocated resources**
2.3.7	TD	**Report results to TACON**
2.4	TD	**Disseminate information**
2.5	FA	**Allocate fighters**
2.5.1	FA	**Conduct on station procedures**
2.5.1.1	FA	**Coordinate take over of service**
2.5.1.2	FA	**Advise WM of system limitations**
2.5.1.3	FA	**Re-arrange comms as necessary**

Task	Team	Communication	Situational Awareness	Decision Making	Mission Analysis	Leadership	Adaptability	Assertiveness	Total Co-ordination
X									
	X	2	3	2	2	3	3	3	2.57
	X	3	3	2	2	2	2	2	2.29
	X	1	3	1	1	1	1	2	1.43
	X	3	2	2	2	2	2	2	2.14
	X	3	3	3	3	2	2	3	2.71
	X	3	3	3	3	3	3	3	3.00
X									
	X	2	2	1	2	2	2	2	1.86
	X	2	3	2	2	3	2	3	2.43
X									
	X	2	3	2	2	3	3	3	2.57
	X	2	3	2	2	3	3	3	2.57
	X	2	3	2	2	2	2	2	2.14
	X	2	3	3	3	3	3	3	2.86
	X	2	3	1	3	3	3	3	2.57
	X	3	3	3	3	3	3	3	3.00
	X	2	3	2	3	3	3	3	2.71
	X	2	3	2	2	3	2	3	2.43
X									
X									
	X	2	3	2	2	3	3	3	2.59
	X	3	3	2	2	3	3	3	2.71
	X	3	3	3	3	3	3	3	3.00

Table 5.4   List of agents involved in the E3D operations

Role of Agent A	Tactical Director (TD)
Role of Agent B	Fighter Allocator (FA)
Role of Agent C	Surveillance Controller (SC)
Role of Agent D	Link Manager (LM)
Role of Agent E	Weapons Controller (WC)
Role of Agent F	Surveillance Operator (SO)
Role of Agent G	Technicians
Role of Agent H	Off aircraft
Role of Agent I	Cockpit

Table 5.5   Matrix showing association between agents on the E3D

	TD	FA	WC	SC	SO	LM	Techs	Cockpit	Off A/C
TD	-	1	1	1	1	1	1	1	1
FA		-	1	1	1	1	1	1	1
WC			-	1	1	1	1	0	1
SC				-	1	1	1	0	1
SO					-	1	1	0	1
LM						-	1	0	1
Techs							-	0	0
Cockpit								-	1
Off Aircraft									-

This is shown in Figure 5.5. Subject Matter Experts (SMEs) considered this to be reasonably representative of the E3D operations. They commented that there would be more communication between the link manager and technicians, between the tactical director and technicians (specifically communications), and between the link manager and external agencies.

The network was used to analyse the centrality of a particular position (positional centrality). This helps to identify the main hub(s) of the social network, and positional centrality can be measured along the 'degree' dimension. Degree is defined as the number of other network positions in direct contact with a given position, and is expressed as a proportion of the total number of nodes in the network. The degree analysis of the E3D operations is presented in Table 5.6. The figures presented in Table 5.6 are simply the proportion of links existing at each node compared to the total number of nodes. The results show that the highest figure for degree is assumed by the Tactical Director, the fighter allocator and the link manager. The Tactical Director and the Fighter Allocator therefore have the maximum number of communication channels open to them (a total of eight, out of a possible eight).

An overall measure of network density can also be derived by dividing the links actually present in the scenario, by all of the available links. For the E3D scenario, the overall network density is calculated as 0.37 (30 links present divided by 81 possible links). This figure is suggestive of a well-distributed (and therefore less dense) network of agents. This figure can also be meaningfully compared among different social networks.

From the generic SNA it can be seen that communications occur between all agents. However to get a better understanding of how many communications occurred between agents individual SNAs were developed. These are detailed in Figures 5.6 to 5.10. The numbers associated with the links between the agents in the

Table 5.6 Comparison for positional centrality (degree dimension)

Agent	Degree
Tactical Director (TD)	0.88
Fighter Allocator (FA)	0.88
Surveillance Controller (SC)	0.77
Link Manager (LM)	0.77
Weapons Controller (WC)	0.77
Surveillance Operator (SO)	0.77
Technicians	0.77
Off aircraft	0.77
Cockpit	0.77

system indicate the strength of association. The strength of association is defined by the number of occasions on which agents exchanged information.

SMEs indicated that this network was reasonably representative. They were surprised that there was communication between the radar technician and the FA and stated that this was unusual (although this might reflect some aspect of the training regime). SMEs commented that the level of communication between OCA and FA was indicative of a new member of staff being trained, as it is far higher than usual. Another comment received was that the LM would rarely communicate with the FA, perhaps only once a mission. They did qualify this by adding the LM would communicate with external agencies a great deal. SMEs quantified this as 50% of

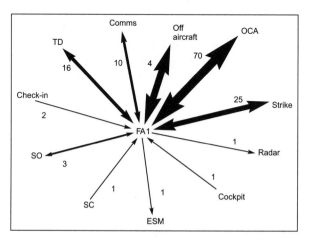

Figure 5.6 FA (1) social network diagram

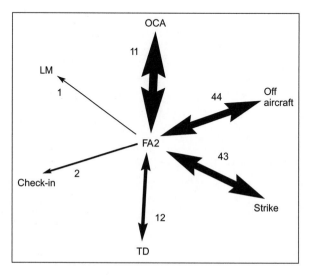

Figure 5.7 FA (2) social network diagram

Case Study in RAF Boeing E3D Sentry   169

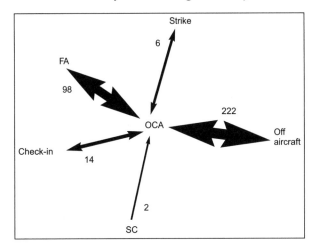

Figure 5.8   OCA social network diagram

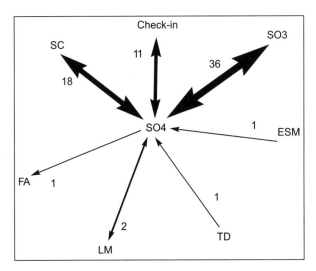

Figure 5.9   SO4 social network diagram

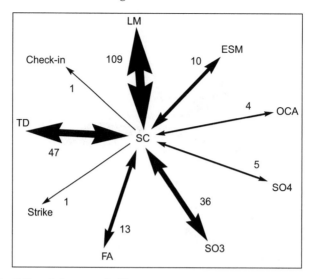

Figure 5.10 SC social network diagram

the communication level between the LM and SC. SMEs commented that this was perhaps missed due to lack of expertise of the observers.

In relation to the level of communication between check-in and FA in comparison to the OCA and strike, SMEs felt that this was reasonable. Check-in may well not communicate very often depending on the mission and they could manage their role without having to interact much internally. SMEs considered this to be an accurate picture of this network.

SMEs commented that the level of communication between check-in and the SO was higher because they were sitting next to one another. It was stated that this would be lower if seated in different positions. This was probably due to the fact that the close proximity allowed communications to be easier. It was observed that a lot of the communications between these two crew members was verbal and showing each other things on the screens. Some of these communications were probably not necessary and would not have happened had the crew members been seated away from each other.

SMEs commented that there would be some communication (around 50% of the communication level between the OCA and SC) between the radar technician and the SC. SMEs also commented that although not represented here, it was important to note that little communication between the SC and ESM takes place.

Operation sequence diagram (OSD)

The operation sequence diagram (OSD) is constructed from the data that were gathered during the observation of the E3D mission. The OSD representation provides a means of summarising the outputs from the CDA and SNA. It represents the tasks, the actors, the communications, the social organisation, the sequence and time in which the scenario took place. The OSD captures the flow of information among distributed actors and shows how this is mediated through technology and team working. The OSDs were constructed for each of the positions observed. The OSD for the Fighter Allocator is shown in Figure 5.11.

Figure 5.11 shows only the functions of 'operations' and 'receive communications' in the OSD. Decisions were made throughout the mission as well as information being transported but because of the nature of the mission it was impossible to note all these down and hence show them pictorially.

Critical decision method

The critical decision method requires a structured interview to elicit responses against defined categories. Unfortunately due to time constraints during the mission it was impossible to conduct interviews with the crew and hence CDMs were not completed.

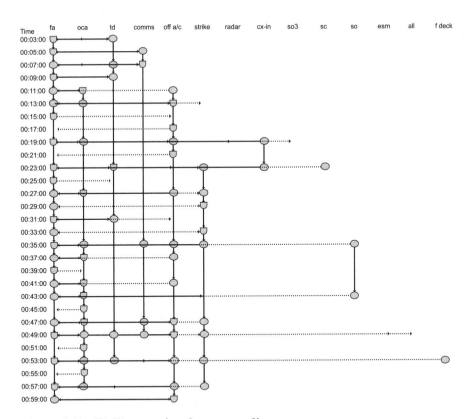

Figure 5.11   FA(1) operational sequence diagram

Propositional Networks

Propositional networks are usually completed using the data collected from the CDMs supplemented by discussion. However, as CDMs were not completed for the E3D mission propositional networks were completed using the HTA. A propositional network shows the knowledge that is related to this mission. The propositional network consists of a set of nodes that represent sources of information, agents, and objects etc. that are linked through specific causal paths. From this network, it should be possible to identify required information and possible options relevant to this mission. The concept behind using a propositional network in this manner is that it represents the 'ideal' collection of knowledge for a mission (and is probably best constructed post-hoc). As the mission unfolds, so participants will have access to more of this knowledge (either through communication with other agents or through recognising changes in the mission status).

Propositional networks were completed for each phase of the HTA and for the overall mission. Figure 5.12 shows the propositional network for the overall mission with Figures 5.13 to 5.20 showing the propositional networks for each phase of the HTA. As was the case for the propositional networks presented in the previous chapter, it is clear how the pattern of knowledge objects changes between the phases.

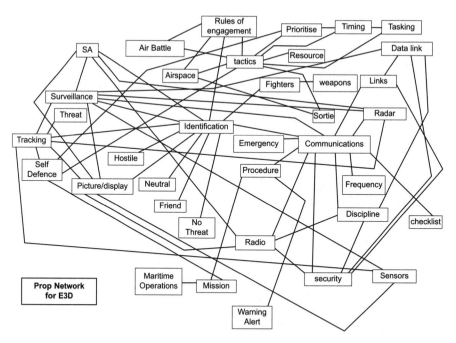

Figure 5.12 Propositional network for the E3D mission

Case Study in RAF Boeing E3D Sentry 173

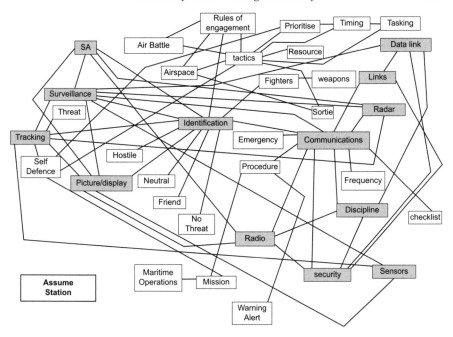

Figure 5.13 Assume station (TD)

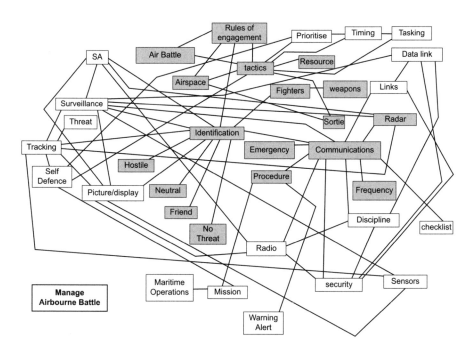

Figure 5.14 Manage airborne battle (FA, WC, TD)

174 Modelling Command and Control

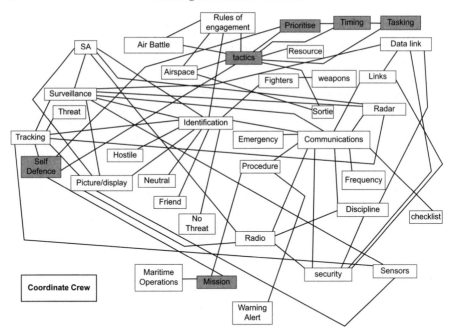

Figure 5.15 Coordinate crew

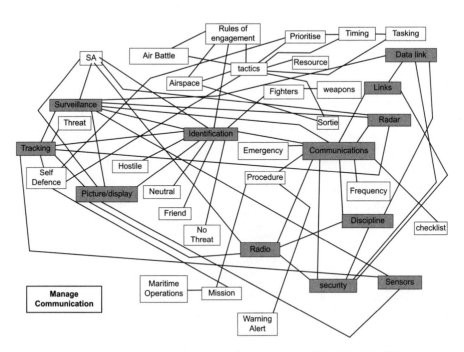

Figure 5.16 Manage communications (SO, LM, TD, CO, CT, SC, RT)

Case Study in RAF Boeing E3D Sentry 175

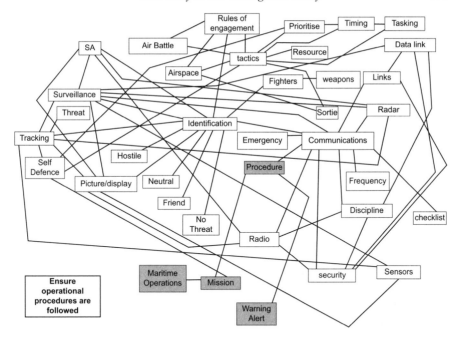

Figure 5.17 Ensure operational procedures are followed (TD)

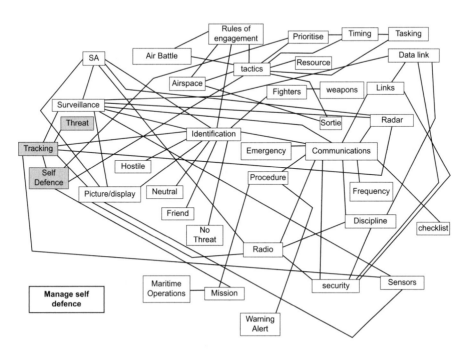

Figure 5.18 Manage self defence (TD)

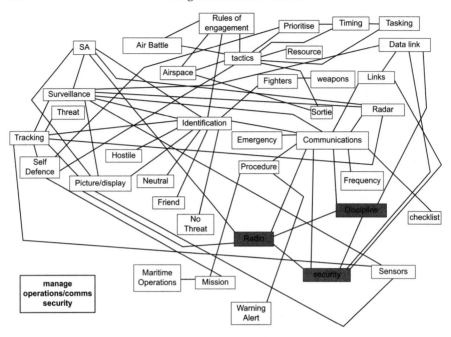

Figure 5.19 Manage operations/comms security (TD)

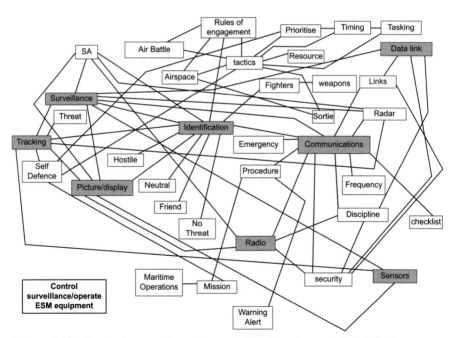

Figure 5.20 Control surveillance/operate ESM equipment (SC, ESM)

Case Study in RAF Boeing E3D Sentry 177

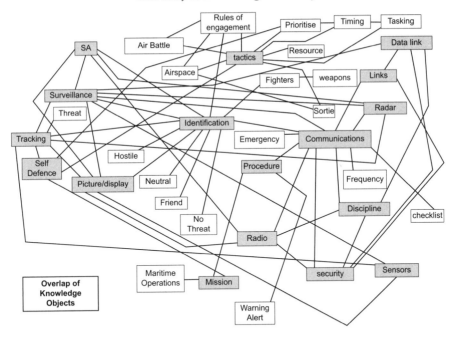

Figure 5.21 Overlap of knowledge objects

The bracketed abbreviations in the figure caption show the crew member(s) responsible for the phase. The majority of the phases are overseen by the TD with just two phases being overseen by other crew members. Figure 5.21 shows the overlap and hence a potential sharing of knowledge objects from each phase. The E3D task has a total of 11 phases of operation, namely assume station, manage airborne battle, coordinate crew, manage communications, ensure operational procedures are followed, manage self defence, manage operations/comms security, and control surveillance/operate ESM equipment. Two of the phases were amalgamated into one for the purpose of the propositional network analysis and another two were single action phases and so were not given a separate propositional network.

Conclusions

In this chapter the various representations are shown and their construction illustrated. One of the issues highlighted by the analysis is the number of safety critical decisions made by crew members over a short period of time. For this scenario, safety critical decisions were made several times in one minute. Another area highlighted is that of the distinction between the quality and quantity of communication. With safety critical decisions being made every few seconds, one might have expected the communications to increase greatly when unexpected circumstances arose. This was not the case observed – in times of high activity the level of communication was much lower than expected. This is reflected in the CDA with communications being registered as medium and low as opposed to high for a great deal of individual tasks.

Coordination demand analysis

According to the CDA, the majority of the E3Ds mission was teamwork (80%). The remaining 20% was defined as taskwork. Coordination between crew members was also fairly high (mean 2.23 out of possible 3). The phases in the overall mission were separated into phases as defined in the HTA. The phase which had the least overall coordination between crew members was 'operate ESM equipment'. This could be due to the fact that few crew members would be involved in this phase and thus little teamwork would be required. The phase that involved the most teamwork was the 'manage self defence'. This would also be expected as it would involve all crew members.

Social network analysis

The social network analysis highlighted three key nodes for this network, those being for the roles of tactical director, fighter allocator and surveillance controller. Meetings with subject matter experts (SMEs) indicated that this was indeed the case and that the measure of competent staff can be made from the levels of their communication. On several occasions, it was stressed that tactical directors should not show a high level of communication as it is their role to supervise, listen and monitor the smooth running of operations. Comments received indicated that should there be an incident the TD will deal directly with those involved and that the SC and FA would deputise for the TD until such time as they are able to return to their supervisory role. It was stated that the TD should be given all necessary information by the crew and thus should not have to ask for it. If the TD has to ask for information, this would indicate that someone has neglected to inform or update the crew.

According to the SNA analysis, the E3D mission was a fairly large but well distributed one, and the TD, FA and LM were the main nodes within the network. The SNA analysis revealed that associations occurred between all agents involved in the scenario. Although it was seen that the FA, TD and LM communicated with all agents the OCA had the most communications with off aircraft (222). Other particularly high communications occurred between the OCA and FA (98) and the LM and SC (109). The amount of communications between the OCA and FA, however, were quite high due to the OCA being in training.

In order to analyse positional centrality in the network, the degree dimension for each agent was also calculated. The TD, FA and LM had the highest degree figure (0.88), whilst the SC, WC, SO, technicians, off aircraft and cockpit had a degree of 0.77. Thus it was concluded that the TD, FA and LM were the main nodes in the social network. A network density figure of 0.37 was also calculated for the E3D mission. This is suggestive of a well distributed network.

One of the main purposes of meeting with SMEs was to evaluate the model of the E3D team that has been made. The issue of meaningful representation was of great concern to both the authors and crew members. By virtue of the fact that CQWI was a training exercise, some recognition that the exercise was simulated rather than real must be made. When asked how the picture of the CQWI training exercise related to a real war scenario, SMEs acknowledged that the picture differs from the real thing, but that a representative picture taken from CQWI would not be too far removed from an actual war scenario. SMEs commented that the training of a new E3D crew member was reflected in the data for the FA. The level of communication between the FA and WC was far higher than normal.

Another issue was how representative the data were on account of observer expertise. Observers did have an opportunity to go on familiarisation flights, which proved to be invaluable. However, both authors and SMEs were concerned that much information was not recorded due to a lack of expertise in listening to multiple channels of information. SMEs commented that mission crew receive intensive training in order to master the skill and that this was not possible to achieve within one or two familiarisation flights. SMEs stated that 'getting back into the swing of things' after a period of leave is even difficult for those who have been listening to multiple channels for years. This was exemplified by the communications missed between the LM and off aircraft. SMEs commented that this was not due to inattention by the observer, merely a lack of expertise and training.

Overall, SMEs stated that the model obtained was reasonably representative. Authors condensed the number of roles in order to build a generic social network from the SOP matrix. This generic SNA shows a wheel like network whereby there are central nodes coordinating the mission. However the CDA shows a fairly high level of teamwork (80%). This indicates that although there are key agents who oversee much of the mission and information flow, the crew as a whole act as a team. The majority of the information is shared between all crew members particularly over the networks and on the screens.

Operation sequence diagram

The OSD presents the activity observed during the scenario. OSDs of the whole scenario could not be constructed they could only be made of the crew members observed. Unfortunately the OSDs constructed only show the transmitting and the receiving of information. Many of the exchanges of information were made by the questioning of and confirmation from the crew, which could not be pictorially represented. Also decisions were being made at such a fast pace that they were impossible to note down. OSDs are useful in that they can show the logical structure of actions involved in a task, however they become problematic when trying to represent large tasks, as is the case here. Despite this limitation the diagrams do show the pattern of communications between particular crew members during the mission.

Propositional network

Propositional networks were constructed to show the knowledge that is related to the E3D mission. Additional propositional networks were completed to show the main phases in the HTA. The individual items are not particular to specific phases of the HTA. Figure 5.21 shows where the phases overlap thus indicating that there is a shared awareness during an E3D mission. About half of the knowledge objects are shared between the phases. As the TD is responsible for the majority of the phases this sharing of knowledge objects makes their tasks easier to complete.

Different core knowledge objects are used at different phases of the mission. Some manipulation is required before these knowledge objects move from one phase to another and some of the communications required to do this have been analysed in the social network analysis and some of the object manipulation has been analysed in the task analysis.

The E3D mission was a highly complex interaction of crew and communication channels. It is an extremely intense environment and over a relatively short period of time (approximately one and a half hours) an enormous amount of communications occurred with information being transferred in seconds. The majority of verbal communications and gestures were recorded, however it should be noted that not all communications could be identified, for example a large amount of information is relayed via the radar screens which will be interpreted by other crew members and acted on.

The physical arrangement of crew members should also be taken into account when looking at the communication networks of the E3D. As mentioned earlier, the majority of the communications occur verbally, either face to face or over the nets, however some are nonverbal. These nonverbal communications are extremely important and rely on the layout of the aircraft. 'Thumbs up' to confirm something is far quicker than engaging with someone over the network and hence the arrangement of crew members is extremely important.

Overall, however a picture was compiled that showed a distributed network of communications, which allows crew members to work fairly independently of each other.

Chapter 6

Case Study in Battle Group HQ

With contributions from Dan Jenkins, Richard McMaster,
Rebecca Stewart, Guy H. Walker and Linda Wells

Introduction

Command and control in the British Army

The British Army's command and control activities are complex and highly evolved. They take place on a number of different levels, each of which is situated within a diverging hierarchy from Division level down to Platoons or Troops comprised of individual soldiers, as illustrated in Figure 6.1. The focus of this chapter is on Battle Group headquarters, relatively close to the theatre of war.

- Command and control activities at this level break down into six distinct yet generic phases:

    1. Set up headquarters and receive direction/orders from higher command formation(s).
    2. Reconnaissance and Planning phase.
    3. Produce and disseminate plans and operational graphics (often only within headquarters).
    4. Wargame.
    5. Execute, monitor, modify and update the plan.
    6. Disassemble command headquarters.

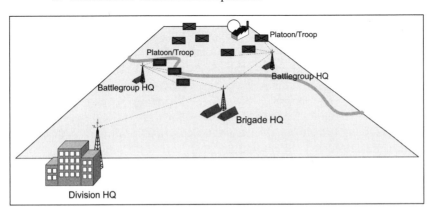

Figure 6.1   Simplified command hierarchy

A number of different tools are currently used to support these six phases. Ordnance Survey maps and clear overlays are used to annotate static features of the map with dynamic aspects of the battlespace. Tables and white boards are also used to represent other planning information. More advanced forms of technology are used that are peripheral to this. They include forms of PC/Laptop computing, advanced radio and satellite communications and various other ISTAR assets and interfaces.

Sources of data

This chapter deals with military command and control activities. Data for this were collected from the following:

- Observation of Command and Staff Training (CAST) exercises at the British Army's Land Warfare Centre in Warminster between 11 to 15 July 2005.
- Observation of military decision making and planning training on 2 to 3 August 2005 at the Land Warfare Centre in Warminster.
- Observation of a Fire Power Demonstration on 11 October 2005 at the British Army's range on Salisbury Plain.

The combat estimate

Information for this and the following sections draws heavily on material used during the experimental team's observation of the combat estimate training noted above. In particular a presentation entitled 'Command and Staff Trainer (South) 1 WFR MiniCAST 2/3 Aug 05' is used with permission. The material is reflected back in order to meet the following objectives:

- To provide the non-military reader with a background to the Combat Estimate technique. This is essential for understanding the EAST analysis that follows.
- To take the opportunity to express the material in psychological or Human Factors terms where this is appropriate.

Background

A process for military command and control is required so that an *adequate* and *flexible* plan is developed in a reasonable amount of *time*. A plan is considered adequate when it meets the commander's intent, provides clear guidance to all sub-units and enough detail to allow the effects of the available combat power to be synchronised at critical points (MoD, 2005b). Flexibility is described in terms of the agility and versatility required to respond to the situation (and enemy) as events occur and change. Timeliness, finally, is about ensuring that there is 'sufficient' time for the battle procedure to be enacted.

The current process used by all players to contribute towards military planning and execution is called the combat estimate. The combat estimate is summed up (and often referred to as) the Seven Questions. These questions break down the process by which plans are made and actions taken; they summarise the activities and outcomes of the different stages of the process. They are as follows:

Question 1:	What is the enemy doing and why?
Question 2:	What have I been told to do and why?
Question 3:	What effects do I want to have on the enemy?
Question 4:	Where can I best accomplish each action/effect?
Question 5:	What resources do I need to accomplish each action/effect?
Question 6:	When and where do the actions take place in relation to each other?
Question 7:	What control measures do I need to impose?

Question 1: What is the enemy doing and why? Question 1 of the Combat Estimate is primarily concerned with developing situation awareness of the physical environment. To this end it produces numerous outputs.

Responding to Question 1 involves a process termed Intelligence Preparation of the Battlefield (IPB). This subsumes two sub-processes; Battlefield Area Evaluation (BAE), which deals with the potential effects of the environment on military operations, and Threat Evaluation, dealing with an assessment of the enemy's capabilities and tactics. Threat Integration is the process of combining the BAE and Threat Evaluation to derive a graphical representation of the situation (a Situation Overlay) as it relates to the static Ordnance Survey map of the battlespace. Figure 6.2 illustrates what happens within Question 1.

The BAE deals with the identification and representation of informational artefacts of interest. These include an analysis of the terrain (physical features such as hills and valleys), mobility corridors/avenues of approach (routes or physical features that lend themselves to the manoeuvring of force elements) and an analysis of the weather and its potential effects. Figures 6.3 and 6.4 illustrate and describe some of the outputs of this process of developing situation awareness about the physical terrain.

The primary concern of the Threat Evaluation phase is to develop situation awareness of the enemy, specifically their tactics and capabilities. There are two main outputs:

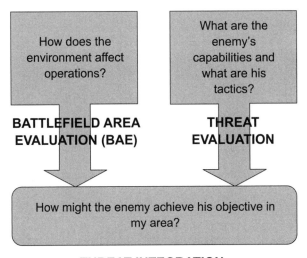

Figure 6.2   IPB system and products (MoD, 2005d)

184 Modelling Command and Control

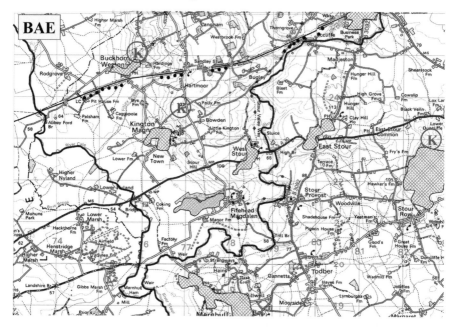

Figure 6.3 Illustration of simplified BAE overlay showing water features and restricted terrain (hashed; MoD, 2005a)

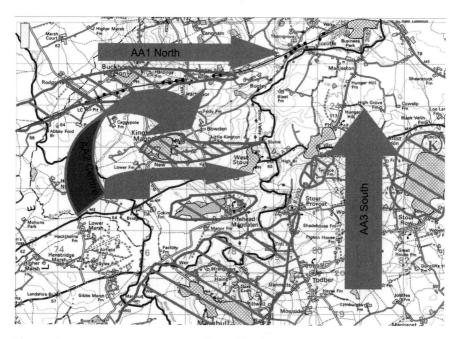

Figure 6.4 Illustration of BAE overlay showing possible avenues of approach to engage enemy force (MoD, 2005a)

- Doctrinal overlays, representing the enemy's doctrinal norms (that is what they normally do in these situations).
- Identification of High Value Targets and vulnerabilities. These are usually an enemy asset or capability that they require for successful completion of their mission (MoD, 2005b). The process of targeting is about 'selecting targets and matching the appropriate weapon system to them' (MoD, 2005b).

The outputs of the BAE and Threat Evaluation phases feed into Threat Integration. 'Threat Integration is the identification and development of likely enemy courses of action' (MoD, 2005b). The outputs of this phase provide, in essence, a representation of the current state of the situation giving particular regard to possible enemy courses of action within it. Figure 6.5 is an illustration of a Situation Overlay and 'provides the [commander] with a clear view of what the enemy might do [...]' (MoD, 2005b).

Question 2: What have I been told to do and why? Question 2 is referred to as Mission Analysis and typically begins with a mission statement. The output of this stage is a Mission Analysis Record. This allows the mission to be analysed in terms of:

- Specified and implied tasks.
- Freedoms and constraints.
- Any deductions made relative to the above.
- Identification of various critical information requirements.

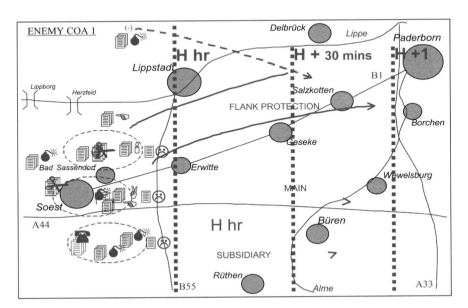

Figure 6.5 **Illustration of a situation overlay; icons show formation of enemy forces, arrows show possible routes/courses of action, dotted lines indicate time phases from the enemy's current position**

Question 3: What effects do I want to have on the enemy and what direction must I give to develop my plan? Question 3 is primarily concerned with what is frequently termed the commander's 'battle winning idea'. A battle winning idea revolves around the concept of 'effects' (on the enemy) and, in Question 3, consideration is given to the 'direction' given to staff in order that these effects are brought about. There are numerous potential effects that can be wrought on an enemy. In offensive operations these might include:

- find
- suppress/fix
- defeat
- strike
- isolate
- secure.

In defensive operations effects might also include:

- deceive
- deny
- disrupt
- block
- turn.

The battle winning idea is expressed in terms of these effects but also with regard to the situational awareness developed in Question 1 combined with cognisance of the overall mission expressed in Question 2. The battle winning idea is expressed graphically in the language of effects using an Effects Schematic, illustrated in Figure 6.6

Question 4: Where can I best accomplish each action/effect? Questions 4 to 7 can be subsumed under the heading 'Course of Action Development'.

Question 4 centres around the creation of a decision support overlay. In practice this entails using the outputs of the earlier BAE (and other materials) to enhance the

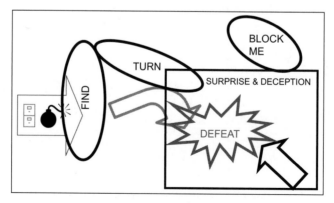

Figure 6.6 **Illustration of the commander's effects schematic; the battle winning idea expressed in terms of effects on the enemy**

Effects Schematic (shown in Figure 6.6). The process normally entails transcribing the commander's effects from the Effects Schematic on to the map, and confirming Named and Target Areas of Interest to produce a Decision Support Overlay (DSO). The DSO is produced at this stage as a draft. It shows aspects of the BAE and the linkage between Named and Target Areas of Interest. Decision points are also added representing where critical decisions need to be taken. Figure 6.7 provides an illustration of a DSO.

Question 5 – What resources do I need to accomplish each action/effect? Question 5 is concerned with the appropriate resourcing of the Named and Target Areas of Interest and the Decision Points elaborated in the Decision Support Overlay. In effect, allocating 'troops to task'.

In this phase the commander is asking what enemy they are likely to see in a particular NAI/TAI, what their likely mission is and what combat power is available. Consideration is given to the outputs of Question 1, in particular the most likely enemy course of action that a response has to be developed to counter for, the outputs of Question 3 and what resources are required to achieve the specified effects. The Decision Support Overlay Matrix (DSO Matrix) is used to achieve these aims. An illustration of a DSO Matrix is shown in Table 6.1.

Table 6.1 shows that TAI number 4 is the first such TAI to be dealt with according to the earlier effects schematic. The purpose of the effect is defined ('Destroy 2xEn Bn Gps sequentially'), the assets associated with this effect are

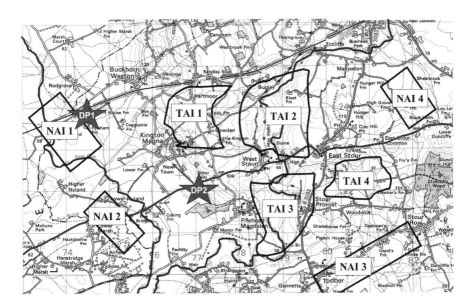

Figure 6.7 Illustration of a draft decision support overlay showing some key features of the BAE in relation to target and named areas of interest (TAI and NAI respectively) and decision points (stars)

listed ('1 x Battlegroup'), the critical information requirements, what NAIs are associated with the TAI and finally, any remaining remarks (for example notes on specifics of formation, number of tanks etc.). Essentially the DSO Matrix is a resourcing tool.

Question 6: When and where do the actions take place in relation to each other? Question 6 is orientated around 'decision points'. These were noted above in Question 4 as they related to events (places). Decision points can also be time based. Decision points (whether event or time based) are key to the timely deployment and synchronisation of forces.

The output of Question 6 is a Decision Support Matrix (DSM). 'The DSM lists the decision points, their locations, the criteria to be evaluated at that point, the actions that are to occur and who is responsible for them' (MoD, 2005b). Table 6.2 illustrates this process with a simplified DSM whereas Figure 6.8 illustrates the same principle with a life-like example.

Question 7: What control measures do I need to impose? 'Control measures are the means by which the commander and his staff coordinate and control what is going on in the area of operations' (MoD, 2005b). This question is typically dealt with on an ongoing basis but is addressed during this phase as a necessary precursor to the Wargaming exercise proper.

For the purposes of illustration, specific control measures might include:

- assembly and engagement areas and limits of exploitation
- fire support
- Nuclear Biological Chemical (NBC) measures.

Control measures also include appropriate combat identification measures in order to reduce the risk of fratricide.

Table 6.1 **Illustration of a partially completed DSO matrix. The matrix lists the order in which TAIs are to be dealt with and the associated resources to be used**

SFR	NAI	TAI	DP	Location	What is the Purpose/ effect	Asset		CCIR	Link	Remark
						Primary	Secondary			
1	-	4	-	Howell Building	Destroy 2 x EnBnGps Sequentially	1 x BG	1 x Sortie Air		NAI1	

Operational graphics

At this point, with the plan in a relatively advanced stage, operational graphics are derived from the Decision Support Overlay. An example is shown in Figure 6.9. Operational graphics are used to disseminate and carry the plan forward into Wargaming and live operations.

Table 6.2 Illustration of a simplified decision support matrix

	Decision Point 1	Decision Point 2	Decision Point 3
Enemy Criteria	IF THE ENEMY ARE DOING THIS…		
Friendly Criteria	AND MY FORCES ARE POSITIONED AND READY…		
Decision	THEN I WILL DECIDE TO DO THIS…		
Where and Who	WITH THESE FORCE ELEMENTS…		
Friendly Actions	AND THESE ARE THE ACTIONS THAT WILL ENSUE.		

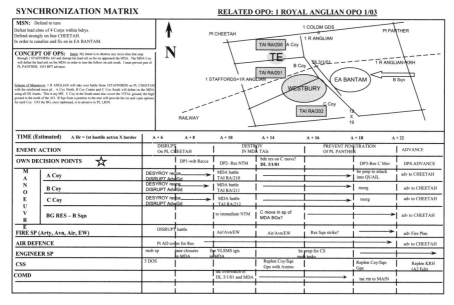

Figure 6.8 Illustration of a completed synchronisation matrix showing the time line (time based decision points) upon which enemy and own courses of action are mapped along with the specific activities of the various force elements

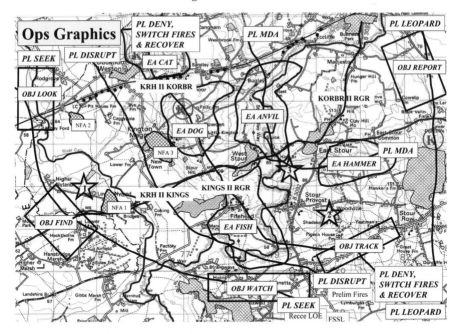

Figure 6.9　Example of operational graphics

Wargaming

Information for this and the following sections also draws heavily on material used during the experimental team's observation of the Combat Estimate training noted above. In particular, a presentation entitled 'Wargaming: Mastering Your Enemy' (based on the Army Field Manual Vol I (Combined Arms Operations) Part 2 (Jul 98) and 3(UK) Div Wargaming Aide Memoir) is used with permission.

Wargaming follows the completion of (at least the first pass through) the Combat Estimate technique and is defined as:

> A process requiring the interaction of the planning staff in a turn based analysis that is used to test a friendly plan (e.g. course of action) against an enemy course of action identified in the Intelligence Preparation of the Battlefield process. [Question 1 of the Combat Estimate above.] (MoD, 2005b)

Wargaming is a visualisation process (not a prediction of the future). It serves to validate the planned course(s) of action against the enemy course(s) of action. It enables anticipatory perspectives to be developed (analogous, perhaps, to the type of 'projection' described by Endsley's (1995) Level 3 situational awareness (SA) and for contingencies to be developed. In the course of Wargaming, the commander has an opportunity to confirm, identify and refine key decisions. It is also a test of the co-ordination and synchronisation of assets.

The Wargame requires members of the planning staff acting in the roles of enemy forces, friendly forces, a recorder (of information) and a referee. Several pieces of planning material are used; the decision support overlay, decision support matrix, and the products of the battlefield area evaluation and intelligence preparation of the battlefield phases. Wargaming relates in the following way to the rest of the Combat Estimate technique (see Figure 6.10).

Execute, monitor and modify the plan

Live operations follow the planning process, the production and dissemination of operational plans and testing of the plan against enemy course(s) of action (Wargaming). Execution involves responding in prescribed ways to orders received from higher command formations as they relate to information derived from the intelligence preparation of the battlefield. The command staff then direct the various force elements to engage the enemy. This is undertaken with voice/radio communications with the planning staff constantly updating dynamic aspects of the battlespace maps, as well as monitoring and where necessary cycling through the Combat Estimate technique to modify the plan.

Summary

It will be evident at this point that a relatively wide ranging analysis is to be undertaken in order to provide a high level overview of Battle group command and control processes. The aim is to place the EAST method under test in a particularly complex domain and to use the results as a basis for any further more detailed insights, perhaps focusing in on particular aspects such as making the plan, Wargaming or executing the plan. The sections that follow describe the application and results of the component methods within EAST.

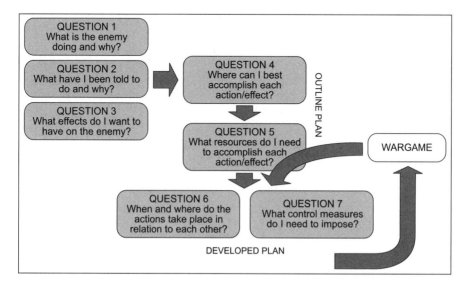

Figure 6.10 Relationship between the combat estimate and Wargaming

Findings

Hierarchical Task Analysis

Task analysis is the activity of collecting, analysing and interpreting data on system performance (Annett and Stanton, 2000; Diaper and Stanton, 2004) and is one of the central underpinning analysis methods within EAST. The first step in the analysis is to conduct an HTA of the total command and control process using the DTC's HTA Tool. The HTA is based on three primary sources of information:

- Direct observation of the scenario/process/work domain as described above.
- A pre-existing task analysis entitled COMBAT Use Case Model produced by General Dynamics UK Ltd (2003).
- Several internal publications and the presentations also described above (MoD, 2005a-d).

Reproduced below is a Task Network (from the HTA) showing the structure in terms of how tasks relate to each other functionally and temporally. The task network is a high level summary representation.

Figure 6.11 shows the interaction between key task elements. The network is a graphical representation of the broadly sequential flow of activities. Setting up the command headquarters is the first step followed by the receipt of orders from which to start developing a plan. Making a plan is comprised of the Combat Estimate 'seven questions' technique. The Combat Estimate cycle sees Questions 1 to 3 feeding into questions 4 then 5, then 6 with 7. The plan is tested by Wargaming, which can provide inputs and modifications to courses of action that then require fresh cycles through the planning process. The production of operational materials and procedures follows, culminating in the phase in which the plan is executed. During the execution of the plan, it is monitored and modified (allowing for the possibility that orders from higher command formations may require changes) in which case the Combat Estimate may again be cycled through. Concurrent with all these activities are tasks related to the display, dissemination and management of information sources. Once the higher commander's mission has been met, and when directed by that formation, the command headquarters will disassemble and possibly step up to the next location to meet the next mission.

Coordination demand analysis (CDA)

It might be assumed that C4i activity will be dominated by coordination activities, but this supposition needs to be checked. Individual tasks from the HTA can be assessed for the type of coordination that is required for successful performance using coordination demand analysis (CDA; Burke, 2005). The method integrates with the HTA, where the tasks identified are assessed according to multi-dimensional aspects of team-working (shown in Table 6.3).

Both task-work (that is task-oriented skills) and teamwork skills (that is behavioural, attitudinal, and cognitive responses needed to coordinate with fellow

Case Study in Battle Group HQ   193

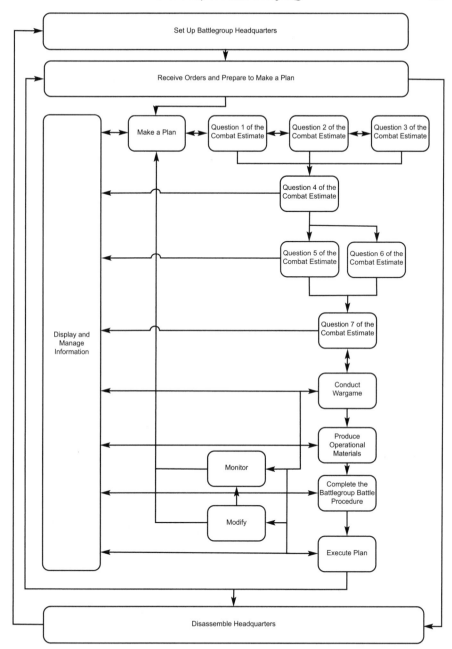

Figure 6.11 Task network of observed military command and control activities derived from HTA

team members) are needed in order to effectively complete team tasks. The CDA procedure allows for the identification of operational skills needed within team tasks, but also the teamwork skills needed for smooth coordination among team members.

Teamwork skills were extracted from the HTA and rated against the CDA taxonomy of communication, situational awareness, decision making, mission analysis, leadership, adaptability and assertiveness (presented in Table 6.3). Each stage of the CDA is scored from 1 to 3 where 1 is low coordination and 3 is high coordination. A score of at least 75% or 2.25 is judged to be relatively high in absolute terms.

From these individual scores a 'total coordination' figure can be derived (based on the mean of the component scores). The results are shown in full in Table 6.4. Key statistics from the CDA analysis show that approximately 48% of the HTA tasks are defined as task work, the remaining 52% as being related to teamwork. Overall, the mean total coordination score is 1.2 (out of a maximum score of three). Mean

Table 6.3   Coordination demand dimensions

Coordination Dimension	Definition
Communication	Includes sending, receiving, and acknowledging information among headquarters staff.
Situational Awareness (SA)	Refers to identifying the source and nature of problems, maintaining an accurate perception of the external operational environment, and detecting situations that require action.
Decision Making (DM)	Includes identifying possible solutions to problems, evaluating the consequences of each alternative, selecting the best alternative, and gathering information needed prior to arriving at a decision.
Mission analysis (MA)	Includes monitoring, allocating, and coordinating the resources of headquarters and battlefield staff, prioritising tasks, setting goals and developing plans to accomplish the goals, creating contingency plans.
Leadership	Refers to directing activities of others, monitoring and assessing the performance of team members, motivation and communicating mission requirements.
Adaptability	Refers to the ability to alter one's course of action as necessary, maintain constructive behaviour under pressure, and adapt to internal or external changes.
Assertiveness	Refers to the willingness to make decisions, demonstrating initiative, and maintaining one's position until convinced otherwise by facts.
Total Coordination	Refers to the overall need for interaction and coordination among headquarters staff

co-ordination scores were also calculated for the seven main stages of the HTA and are shown in Table 6.5. The majority of the coordination scores are low, showing that there is a low level of teamwork occurring. The phase which had the least overall coordination between actors in the scenario was 'Translate Products of Q1-7 into Operational Graphics' and 'Select Battle Procedure'. This could be due to the fact that few personnel would be involved in these phases and thus little teamwork would be required. The phase that involved the most teamwork was 'execute plan'. This would also be expected as it would involve all planning personnel.

Comms Usage Diagram

The communications technology observed in the command and control scenarios was as follows:

- Voice communications in the form of aural communication between individuals and louder global announcements within the command tent.
- Radio communications between personnel in the command tent and individuals in the battlespace.
- Operational graphics and other visual planning aids such as clear overlays and whiteboards.

Current army command and control activities do not necessarily have a particularly complex underpinning communications infrastructure in place (compared, perhaps, to various instantiations of NEC such as BOWMAN or civilian forms of C4i such as air traffic control). Advantages in terms of robustness are, however, noted.

Communications within the command tent are conducted principally by voice. The planning process involves verbal dialogues between planning personnel in close proximity (during meetings) and announcements where the command staff will shout (for example timescales for upcoming meetings or deadlines). Radio communications are also an integral part of the command process. Radio communications take the

Table 6.4   CAST scenario CDA results

Category	Result
Total task steps	132
Total task work	64 (48%)
Total teamwork	68 (52%)
Mean Total Co-ordination	1.2
Modal Total Co-ordination	1.1
Minimum Co-ordination	0.3
Maximum Co-ordination	2.6

Table 6.5   CAST scenario CDA results in HTA stages

Category	Prepare Plan	Display and Manage Information	Make Plan	Translate Products of Q1-7 into Operational Graphics	Conduct Wargame	Select Battle Procedure	Execute Plan
Mean Comms	2.5	2	2.2		2		3
Mean SA	2	2	2.2		2		2
Mean DM		1.5	1.8		1		2
Mean MA	1	2	2.1		1		1
Mean Leadership	1.25	1.3	1.6		2		3
Mean Adaptability		2	2		3		2
Mean Assertiveness			1.8		1		2

form of verbal dialogues using a standardised radio telephony method. Radio communications are intercepted by individual members of the planning staff (they cannot usually be heard by all personnel in the command tent). The final group of communications media are the operational materials produced out of the planning process along with the aids such as whiteboards and clear overlays that are used during it.

Figure 6.12 provides an illustration of the broad context within which verbal communication takes place, while the figures contained within the Combat Estimate section illustrate some of the planning material used. All of these methods of communication are to be analysed using the CUD method (Watts and Monk, 2000)

Advantages and disadvantages of existing comms media
CUD is a structured and exhaustive method, the first stage in which is to define the communications media and to tabulate the advantages and disadvantages associated with each (Watts and Monk 2000). Table 6.6 presents the outputs of this analysis.

Critique of comms usage
The last stage in the CUD method is to provide a critique of existing communications technology based on the task flow, the identified advantages and disadvantages, and also the 'knowledge objects' that require sharing between agents in a given task step (see later section on propositional networks). A number of possible alternative technical solutions help to shape thinking in this respect and the following evaluation

Figure 6.12   Photograph of the command tent

of comms usage results from this assessment. It should be added that this critique is not indicative of any actual or proposed recommendation as such, rather a consideration of possible alternatives and issues based on the data collected.

Verbal communications
A compelling advantage of verbal communications is the immediacy and error recovery opportunities provided. In addition the role of military rank imposes a high degree of structure and authority within the decision making process. There remains, however, the possibility of psychological issues such as 'group think' and bias in using this form of communication in this context, although there are techniques available to help overcome this (Janis, 1982). Based on our observations, high ambient noise levels were noted to the extent that the intelligibility of verbal communications could easily be affected. Aside from these issues the immediacy of verbal communications appears well matched to the dynamic nature of the scenario.

NEC solutions
The Intelligence Preparation of the Battlespace requires a number of transformations to be undertaken to represent the data. Cognitive effort is, therefore, required to achieve adequate levels of situational awareness from which to develop and resource courses of action. Because of the prominent cognitive component, the potential exists for ambiguities and failures in shared understanding and SA to disrupt the process of decision making. NEC approaches that embody positional data, 3D representations

Table 6.6   Advantages and disadvantages of existing comms media

Media	Advantages	Disadvantages
In-Person Voice	Physical verification that correct individual (and planning role) is being referred to. Favourable role of non-verbal communications in aiding shared understanding. Possible favourable role of military rank in face to face communications. Sharing of explanatory resources such as operational graphics/whiteboards etc.	Possible detrimental role of social status/military rank. Possible contextually related distractions (e.g. noise and general confusion in the planning tent). Possible ambiguity in physically pointing out and referring to shared resources (e.g. plans and whiteboards). Relatively static descriptions of a highly dynamic and spatially dispersed scenario.
Radio Communications	Sound stable. Possible for communications to be recorded for post-hoc analysis and training. Hands free. Time saving with common abbreviations and nomenclature. Enhanced intelligibility with common abbreviations and nomenclature. Read-back provides validation of shared understanding. Open channel radio comms. aids shared SA among other units and members of planning staff. Favourable implications of military rank could be diluted with distance and standardised radio protocol.	Intelligibility can be an issue with distortion/artefacts in radio comms. Language/accent ambiguities. Unscheduled/ad-hoc presentation of comms. Any informality/abbreviation in comms relies on assumption of shared meaning. Relatively slow communications compared to other comms solutions. Translation from verbal domain to visio spatial domain required (and vice versa). Open channel radio comms could permit simultaneous comms on same frequency causing masking. Read-back can be out of synchronisation with current activities if sender/recipient are slow to respond. Unfavourable dilution of favourable aspects of military rank.
Operational Graphics and Planning Aids	Paper based materials can be substantially degraded without information loss. Relatively easy to derive with little extra training required in having to use a pen and paper.	Static representations of typically dynamic scenarios. Training load relatively high in the use of methods to overcome the disadvantages of representing dynamic 3D phenomena as 2D paper based representations. Complexity and dynamism of scenario can cause administrative bottlenecks. Legibility of graphics and handwriting. Whiteboards and overlays potentially cumbersome to handle. Document tracking and administration potentially difficult.

of the battlespace and live updating of it directly from the field promise significant advantages. Specific advantages include efficiency gains and for the commander to be able to rapidly orientate and couple themselves to the dynamics of the situation.

Summary

The main advantage of the CUD method is that it provides a detailed analysis of the communications media used in the scenario, which in turn relates to the communications links extant between people (and technology). These links are represented in the Social Network Analysis that follows. In other words, when other factors are considered, the advantages and disadvantages associated with each communications modality may represent an enhancing or constraining feature (depending upon how the network of people and technology is configured and how it is put to work).

Social network analysis

Social network analysis (SNA) is a means to present and describe the underlying network structure of individuals or teams who are linked through communications (Driskell and Mullen, 2005). The social network diagrams that follow represent people and technology in the scenario, and the communication links (specified by the CUD method above) that exist between them. The main benefit of the SNA method is that it is able to represent not just human 'agents' in the scenario, but technological ones as well. The criterion is that each 'node' in the network has the ability to communicate and/or transform information. A node, therefore, is any person or item of technology represented in the network; an agent is a human actor only.

The relationships that are specified from this analysis can be used to determine what aspects of the network structure constrain or enhance the performance of nodes in the network (Driskell and Mullen, 2005). For the purposes of this analysis, a high level systems view is taken. Key nodes are identified and can, in some cases, be grouped together according to a common role in the scenario. This is illustrated in the social network diagram shown in Figure 6.13.

Description

The current social network defines eight key nodes, some of which are comprised of further sub-systems (as illustrated). The nodes include the Higher Command Formation, the Commanding Officer (CO), Chief of Staff (COS/2IC), the 'Principal' Planning Staff such as the IO/G2 (to varying extents it also requires the participation of individual roles such as Recce/ISTAR, Eng, A2/Log and Arty/AD. These have been placed on the periphery of the planning/principal staff node for illustration. There are also other ancillary command staff (such as those responsible for more general tasks and information management), what are referred to as sub-units (who are typically carrying out activities live in the battlespace) and, finally, the collection of graphics and planning aids derived from the Combat Estimate (artefacts that represent and transform information in some manner).

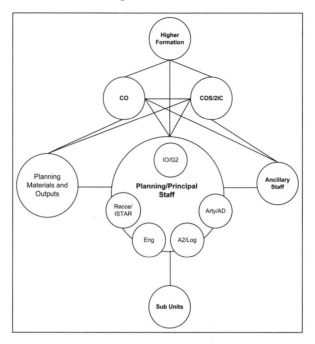

Figure 6.13 Social network diagram illustrating a systems level view of the combat estimate scenario

Activity stereotypes

The social network is dynamic and different nodes and links become active under different activity stereotypes. These are revealed by detailed analysis of the HTA. The activity stereotypes are as follows:

- Briefing or providing direction: here the Commanding Officer (CO) is directing communications and information outwards to subordinate staff in a prescribed and tightly coupled manner (particularly Questions 1 and 3 of the Combat Estimate).
- Reviewing: here the planning/principal staff communicate in a more collaborative manner with mutual exchange of information and ad-hoc usage of planning materials and outputs (in particular Questions 2 and 5).
- Semi-autonomous working: here members of the headquarters staff are working individually on assigned tasks and become relatively loosely coupled in terms of communication. The communication channels remain open but used in an ad-hoc, un-prescribed manner (occurs at various points in all Questions).

Calculation of social network metrics

Network metrics enable the visual representation of nodes and links to be presented according to a number of different dimensions using numerical indices. The dimensions to be assessed in this case are 'centrality' and 'network density'.

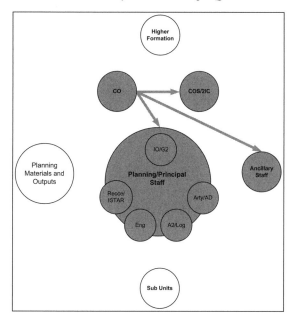

Figure 6.14 'Briefing': illustration of social activity and communication. The one way flow of information from commanding officer to subordinate staff is evident as is the close coupling between nodes

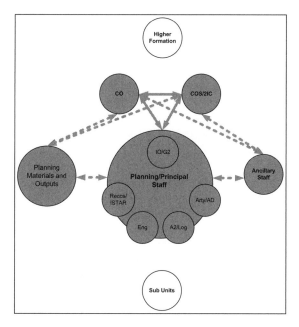

Figure 6.15 'Reviewing': illustration of social activity and communication. The collaborative, 2-way nature of communication is evident as are the ad-hoc 'open' links to planning materials (shown by dotted links)

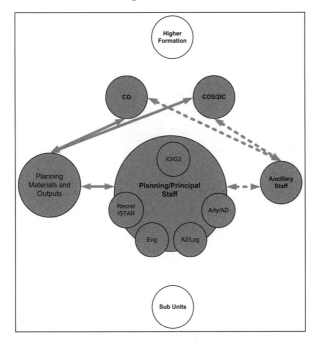

Figure 6.16 'Semi-Autonomous working': illustration of social activity and communication. The agents are primarily linked to the planning materials that are being worked upon with 'ad-hoc' open channels to other Headquarters staff (shown by the dotted lines)

Centrality mathematically defines 'key nodes' in the scenario, based primarily on the number of links. Density is a representation of the interconnectivity of a network. Both parameters can be discussed within a frame of reference that allows judgements to be made as to the efficacy of the network in terms of whether key agents/centrality and density may influence system performance. Taken as a whole, the network yields the following results (Table 6.7).

The planning/principal staff followed by the Commanding Officer and Chief of Staff/2IC emerge as key nodes overall. The density metric also reveals a relatively highly interconnected network, especially compared to alternate domains where figures of around 0.2 are common.

Table 6.8 presents the results of the same analysis showing how these parameters alter during each of the three identified activity stereotypes (briefing, reviewing and semi autonomous working). The degree of agent centrality does not appear to alter to a great degree (the Commanding Officer, COS/2IC and Planning/Principal staff remain as key nodes) except in the case of Semi-Autonomous Working where the role of the various planning materials becomes more prominent (all agents appear to link with them when engaged in this mode of working). Network density does change markedly. These results reflect the manner in which the communications network 'reconfigures' itself according to the activity stereotypes being undertaken. For example, the 'one way' flow of information during the briefing phase is indicative of less interconnectivity (it is only the Commanding Officer who is linked to subordinate staff), whereas the more collaborative 'reviewing' activity stereotype is more densely configured, as would be expected with two way links between most or all nodes.

Table 6.7 Network metrics illustrating centrality (key agents in the scenario) and density (network connectivity) for the social network as a whole

Agent	Agent Centrality
Higher Formation	0.89
Commanding Officer	1.11
COS/2IC	1.11
Ancillary Staff	0.67
Planning/Principal Staff	1.33
Sub Units	0.22
Planning Materials and Outputs	0.67
NETWORK DENSITY	0.31

Table 6.8 Network metrics illustrating centrality (key agents in the scenario) and density (network connectivity) for the activity stereotypes of Briefing, Reviewing and Semi-Autonomous Working

	Centrality		
Agent	Briefing	Reviewing	Semi-Autonomous
Higher Formation			
Commanding Officer	0.33	0.67	0.33
COS/2IC	0.11	0.67	0.33
Ancillary Staff	0.11	0.33	0.33
Planning/Principal Staff	0.11	0.67	0.33
Sub Units			
Planning Materials		0.33	0.67
NETWORK DENSITY	0.03	0.20	0.13

Facilitation of network links

Linking in with the previous results provided by the CUD method, Figure 6.17 illustrates the communications media that facilitate the links between nodes in the network. These results are also summarised in Table 6.9 as a communications/modality/technology matrix. At this level of analysis it is apparent that the communications infrastructure is relatively basic, being comprised of paper media and verbal communications either in person or mediated by radio. The advantages of a process based on this level of facilitation is of course its robustness, but clearly there are opportunities to, for example, more rapidly acquire the state of situational awareness through novel technology that does not necessarily rely on verbal communications and manual updating of maps.

Operation sequence diagram (OSD)

The OSD is the main summary representational method within EAST. It does not necessarily add new data, rather it unifies and summarises the previous analysis. Process charts offer a systematic approach to describing activities. Symbols are used to denote a broad class of activity (such as performing an 'operation', 'receiving'

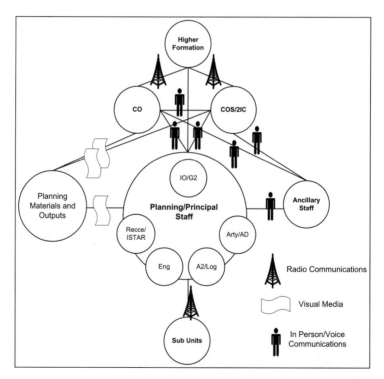

Figure 6.17 **Illustration of social network where the communication links between nodes have been annotated with the media or modality that facilitates it. The sub-units and higher formation are geographically remote from the Battlegroup Headquarters**

Table 6.9 Technology/facilitation/modality matrix. Shading shows the match between technology and modality

Modality	Technology/Facilitation		
	Radio	Planning Aids	In-Person Voice
Verbal	■		■
Visual		■	■
Written		■	

information, or 'decision' making). These symbols refer to the player's (and tasks that they are performing) in the scenario, how they are linked together, and how they map onto the scenario timeline. Information for each activity/process is drawn directly from the scenario HTA. The number contained within the symbol refers to the relevant task step in the HTA.

Charting techniques represent the temporal structure and interrelations between and among activities and players. The OSD, therefore, attempts to emphasise key features using a graphical representation that is easy to follow and understand (Kirwan and Ainsworth, 1992). A further benefit of this form of representation is that fact that it preserves the ability of preceding methods, such as SNA and HTA, to represent human and non-human elements of the system.

Several enhancements to the OSD representation have been made to reflect the outputs of supporting analysis methods contained within EAST. The OSD presents a temporal overview of tasks, the outputs of the CDA, communications media from the CUD and links between agents from the SNA. The approach, therefore, summarises a large amount of supporting analysis in a fashion that is graphical and relatively easy to follow, whilst also being scaleable to suit different 'sized' scenarios. This enhanced version of OSD is one of the key summary representations within the EAST methodology. The figures that follow present the complete OSD for this scenario. To aid in the interpretation of the diagrams a key is provided in Figure 6.18 and a short narrative is provided for each subsequent diagram.

Receiving orders (Figure 6.19)
This OSD chart illustrates the phase preparatory to making a plan. It focuses on the receipt of orders from higher command formations and the dissemination of those amongst the local headquarters. (It can be noted that the Higher Formation is performing an operation [issuing orders], represented by the circular symbol that is being received by the Commanding Officer, represented by the rounded box symbol).

Gathering information (Figure 6.20)
This OSD chart illustrates question 1 of the Combat Estimate (what is the enemy doing and why).

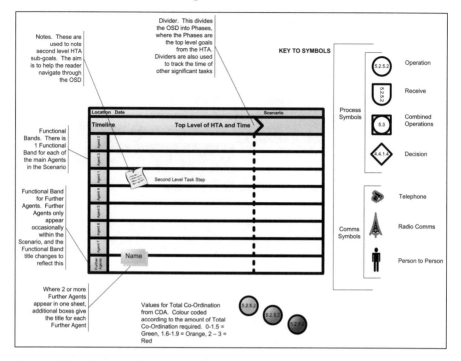

Figure 6.18 Key to enhanced OSD symbology

Information is flowing inwards from the field agent(s) via radio and is disseminated amongst key personnel. Two recurring patterns are evident, firstly the briefing provided by the commander. This task occurs concurrently throughout.

Confirmation of orders (Figure 6.21)
This OSD chart illustrates questions 2 and 3 of the Combat Estimate. (What have I been told to do and why? What effects do I want to have on the enemy and what direction must I give to develop my plan?)

It begins with the remotely mediated confirmation of orders followed by the reviewing phases mentioned earlier (where more collaborative working is in evidence). Note that question 3 commences with a defined decision making component.

Decision making and planning (Figures 6.22 and 6.23)
These OSDs illustrate questions 4, 5, 6 and 7 of the Combat Estimate. (Where can I best accomplish each action/effect? What resources do I need to accomplish each action/effect? When and where do the actions take place in relation to each other? What control measures do I need to impose?)

Once again the presence of task step 2 is noted (management of information). It is also possible to discern a densely interconnected web of operations involving the principal planning staff. Questions 5 and 7 appear to be concerned with a great deal

Case Study in Battle Group HQ   207

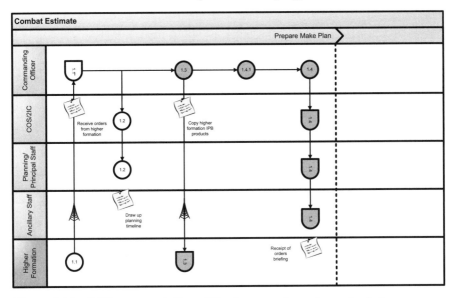

Figure 6.19 OSD representation of the 'prepare to make a plan' phase

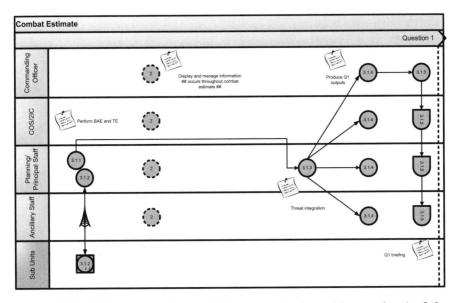

Figure 6.20 OSD representation of the phase dealing with question 1 of the Combat Estimate

208 *Modelling Command and Control*

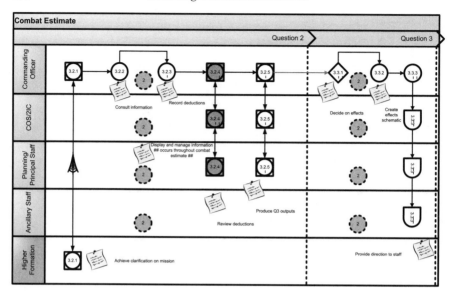

Figure 6.21 OSD representation of the phase dealing with questions 2 and 3 of the Combat Estimate

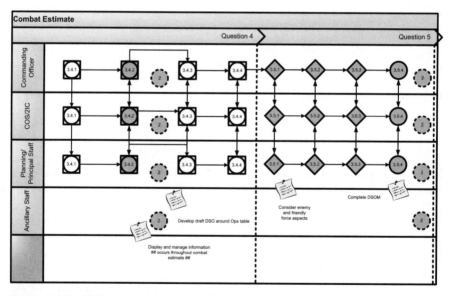

Figure 6.22 OSD representation of the phase dealing with questions 4 and 5 of the Combat Estimate

of decision making activity. Overall a high degree of collaborative working is in evidence in the latter phases of the Combat Estimate.

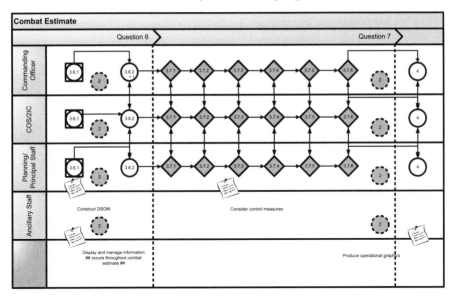

Figure 6.23 OSD representation of the phase dealing with questions 6 and 7 of the Combat Estimate

Execute the plan (Figure 6.24)
This OSD presents the concluding phases of the planning process and the beginning of the execution of the plan. The execute plan phase is presented merely in summary but in outline form the generally vertical disposition of operations can be noted.

In summary, the OSD illustrates many of the facets dealt with in earlier methods. Team working and coordination is reflected in the varying patterns of connectivity between and among operations, and the presence of significant team working and communications related operations. The social network stereotypes (reviewing and briefing) are also represented, and the changing pattern of links and operations reflects this. Overall, it appears that the planning and execution process moves through three distinct phases. In the earlier phases there is a focus on the hierarchical/vertical flow of information. In the second phase the operations (and agents) become closely interconnected with decision making components dominating. In the third and final phase the pattern of operations once again assumes a vertical/hierarchical disposition. At the highest level, then, the command and control process assumes a pattern of information retrieval, closely knit decision making processes and then dispersion and action.

Propositional networks (and the critical decision method)

In recent years the study of decision making in real-world situations has received a great deal of attention, with a growing emphasis on the use of interviews to collect such information. The critical decision method (CDM; Klein and Armstrong, 2005) is a contemporary example. According to Klein, 'The CDM is a retrospective interview strategy that applies a set of cognitive probes to actual non-routine incidents that

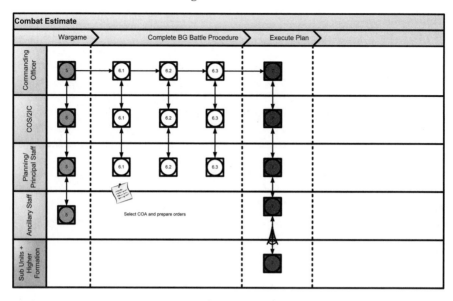

Figure 6.24 OSD representation of the concluding phases of the Combat Estimate and execution of the plan

required expert judgment or decision making' (Klein, Calderwood and MacGregor, 1989, p.464). In this approach, the interview proceeds through four stages: briefing and initial recall of incidents, identifying decision points in a specific incident, probing the decision points, and checking. A slightly modified version of the CDM probes presented in O'Hare et al. (2000) are adopted within EAST, as described previously in Table 3.9 (see Chapter 3).

In the present context this procedure permits elicitation of information on key decision points as well as non-routine 'incidents' and feeds directly into the construction of the propositional networks that follow. The raw transcripts derived from the respondent's answers to the probe questions are subject to a content analysis to extract 'knowledge objects'. Knowledge objects are defined as an 'entity or phenomenon about which an individual requires information in order to act effectively' (Stanton et al., 2006). Nouns such as 'effects', 'radio', 'map' and 'weather' are usually good candidates for knowledge objects, although there are many other knowledge based facets of a scenario revealed by the CDM.

From the CDM it is possible to construct Propositional Networks to show the knowledge that is related to each scenario. The propositional network consists of the set of knowledge objects extracted earlier linked through specific causal paths. For example, 'courses of action' has the property of 'friendly forces' and 'uncertainty', and so on. The deeper, more fundamental concept that this collection of linked knowledge refers to is situational awareness. From the resulting network it is possible to identify:

- The structure and temporal nature of distributed SA.
- The knowledge underpinning decision making.

The concept behind using a propositional network in this manner is that it represents the 'ideal' collection of knowledge for an incident (and is probably best constructed post-hoc). As the incident unfolds, so participants will have access to more of this knowledge (either through communication with other agents or through recognising changes in the incident status). Consequently, within this propositional network SA can be represented as the change in weighting of links as well as the total quantity of knowledge objects related to the scenario.

Both the OSD and the PN are representational methods within EAST and enable the distributed and dynamic nature of the scenarios to be captured. In Figure 6.25 the complete assemblage of knowledge (for the entire system/scenario), and their various interactions, are presented. In successive diagrams (Figures 6.26 to 6.31) the change in activation of this knowledge contingent upon task phase is also presented (in which shading indicates that a node is currently active). Although these figures are quite small, it is clear how the pattern of knowledge objects changes between the phases.

The summary Table (Table 6.10) presents an analysis of core knowledge objects for each phase in the scenario; in other words, what are the essential pieces of knowledge relating to each phase in the scenario. Core knowledge objects are defined by network metrics (similar to those used earlier in the SNAs). Core knowledge objects also feed backup to the CUD method earlier, where they help to improve the level of critical analysis carried out on the communications technology in use (that

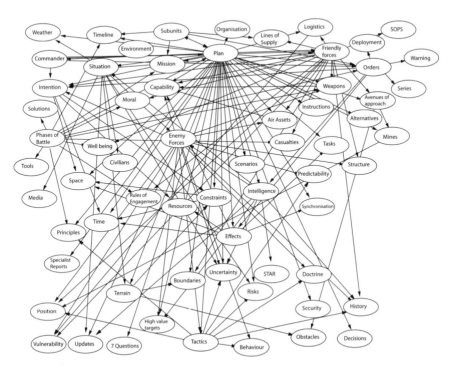

Figure 6.25 Overview of systems level knowledge for the CAST scenario represented via propositional network

212   *Modelling Command and Control*

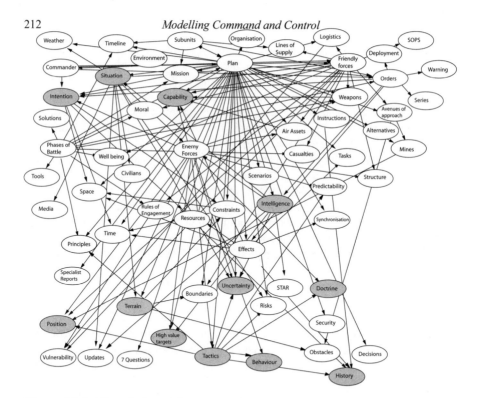

Figure 6.26 Knowledge objects associated with question 1 of the Combat Estimate technique

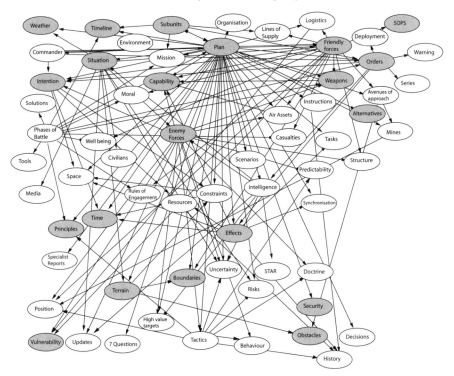

Figure 6.27 Knowledge objects associated with questions 2 and 3 of the Combat Estimate technique

214 Modelling Command and Control

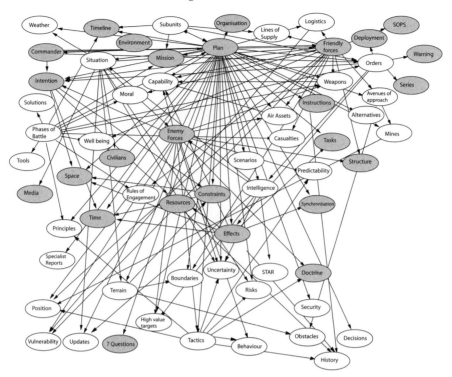

Figure 6.28 Knowledge objects associated with question 4 of the Combat Estimate technique

Case Study in Battle Group HQ 215

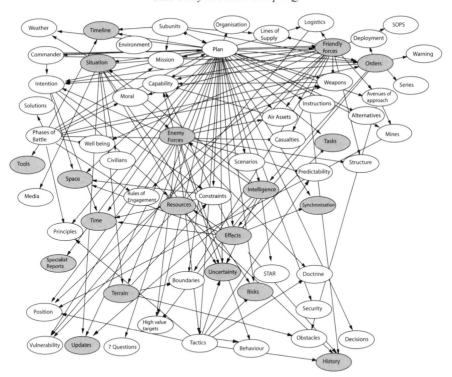

Figure 6.29 Knowledge objects associated with question 5 of the Combat Estimate technique

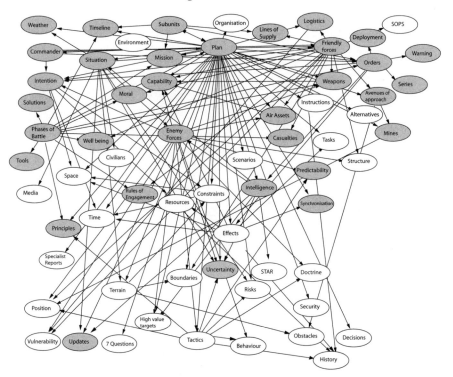

Figure 6.30 Knowledge objects associated with questions 6 and 7 of the Combat Estimate technique

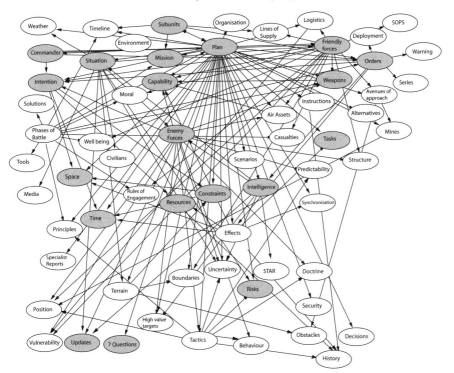

Figure 6.31 Knowledge object activities with putting the plan into effect

is exactly what knowledge is the communications technology/media in the scenario trying to facilitate; can it be facilitated in more optimal ways?).

There are 18 knowledge objects that, according to the criteria of 'five links or greater', can be referred to as key. As the Combat Estimate process progresses through its distinct phases it can clearly be seen that the activation of these key objects changes in type and structure (Table 6.10). Uncertainty, terrain, position and tactics predominate in the early phase of the process, whereas intelligence and courses of action dominate later phases (with friendly forces, situation and enemy

Table 6.10   Summary of key knowledge objects active within each scenario

Key Knowledge Objects	Combat Estimate Q1	Combat Estimate Q2 and Q3	Combat Estimate Q4	Combat Estimate Q5	Combat Estimate Q6	Implementing Plan
Subunits		■			■	■
Plan		■			■	■
Friendly forces		■		■	■	■
Orders		■			■	■
Situation	■	■		■	■	■
Intention	■	■	■		■	
Capability	■	■	■		■	■
Phases of battle	■	■			■	
Weapons		■	■		■	
Enemy			■	■	■	■
Intelligence	■		■	■	■	■
Effects			■	■	■	■
Courses of action				■	■	■
Uncertainty	■	■				
Terrain	■	■		■		
Position	■	■				
Tactics	■					
History				■		
Total Knowledge Objects	**9**	**12**	**6**	**8**	**13**	**10**

dominant throughout). The advantage of this approach, certainly at this high level of analysis, is that it pinpoints the changing sources of information that actors in the scenario draw upon to develop situation awareness, and may well be a useful approach to defining training needs.

Conclusions

The aim of this chapter was to describe the application of the EAST methodology to a military command and control scenario and to demonstrate its capabilities in terms of revealing the emergent properties of it. This is the first time that the EAST method has been applied in the military arena. There follows a short summary of the main findings, a comparison with EAST analyses conducted in alternate C2 scenarios where NEC is already extant, and a short section on future directions.

This chapter is couched at a summary level of analysis and based upon observation of one particular military scenario. The results require interpretation with those caveats in place, but in so far as generalisations and characterisations can be made, the following emergent properties and issues have arisen from the EAST analysis:

1. Command and control relies heavily on tasks that require interaction with other team members, and where this is manifest, team working is principally concerned with the communication of information and development of SA.
2. A relatively simple, yet robust, technological infrastructure underpins team tasks. It is heavily reliant on a combination of verbal communications and/or the translation of various planning 'products' into an integrated, collective, 4D spatial and temporal 'image' of the battlespace. It appears to be in this domain, based on the CUD method, that NEC technology has much to offer. The assumption is that if the state of SA can be more rapidly and accurately acquired (and there seems little doubt that new technology offers this potential), then decision superiority can be achieved more quickly. If SA can also be shared in optimal ways throughout the system (which again, new technology appears to provide for), then unity of effort can also be achieved.
3. The task analysis specifies how the configuration of people and technology changes in a task and context dependant manner. Three activity stereotypes are defined, semi-autonomous working, briefing and reviewing. The social network configures (and re-configures) itself numerous times during the enactment of military command and control (and the Combat Estimate specifically). As the network is re-configured, the constraints of it in terms of communications, density and centrality change. The design of NEC paradigms, therefore, is revealed to be more than just a consideration of technology in isolation. The specification of technology may be appropriate for one configuration, but inappropriate for another. The combination of the task analysis and SNA appears to provide one route into addressing this issue.
4. The knowledge base that underpins effective SA at the systems level changes in response to task phase, but also arises as a property of the constraining features of the configuration of people and technology. Systems level SA, at

this summary level of analysis, appears to support, and be congruent with, task goals.

In summary, the emergent properties associated with military (and indeed any) command and control scenario relate to the interplay between task, social and propositional networks. Further insights are gained from more direct comparisons with civilian examples.

Chapter 7
Development of a Generic Process Model of Command and Control

With contributions from Rob Houghton, Dan Jenkins, Richard McMaster, Paul S. Salmon, Rebecca Stewart, Guy H. Walker and Mark Young

Three Domains for Command and Control

The bases for the development of the generic model are field studies conducted in three domains. These were the emergency services (Police and Fire Service – McMaster, Baber and Houghton, 2005; Houghton, Baber and McMaster et al., 2006), civilian services (National Grid, National Air Traffic Services, and Network Rail – Walker et al., 2006b) and armed services (Air Force E3D, Navy type 23 frigate and Army CAST brigade level exercise). The latter studies conducted in the armed services are presented in Chapters 4, 5 and 6 of this book. It is the cumulative understanding of command and control, developed through a variety of domains, which led to the development of a generic model.

Emergency services

Two command and control application areas were analysed in the emergency services: the fire service and the police. In the UK (and several other countries across Europe), the emergency services operate a tripartite control structure. Major incidents, which require high-level, strategic command, are termed Gold. Typically, these occur when the co-ordination of a great many units is required. Usually such command is not required and command can be exercised on a local, operational level, which characterises Bronze command. Between Gold and Bronze lies a tactical command level termed Silver.

Fire service
In order to study fire operations, observations were conducted at the Fire Service Training College. The training college provided access to command structures for ecologically valid exercises, without the potential risk associated with actual fires. During each exercise, an incident commander (the Assistant Divisional Officer or ADO) issues commands to the sectors being controlled. For a medium-to-large incident, within the Silver command level it is necessary to divide response into sectors, which can be either geographical (that is parts of the fire ground), or functional (for example, managing water supplies or a breath-apparatus crew). The

exercises that were observed covered operations including the search for a hazardous chemical, fire in a chemical plant, and a road traffic accident.

Police

Police operations were studied through observations in Force Command and Control (FCC) and Operational Command Unit (OCU) sites. The focus was on accidents requiring immediate response, that is, where suspects were on the premises or an incident was in progress. FCC Emergency Call Operators prioritise incidents as requiring immediate, early or routine response, according to their urgency. Incidents that are graded as 'Immediate Response' are those that require an urgent Police presence, usually because there is a high risk of serious injury or death, or where there is a good chance of an arrest if the response is rapid (that is when the crime is still taking place). When an incident is prioritised 'Immediate Response', only the bare minimum of details are taken from the caller by the Emergency Call Operator (location, nature of emergency and caller's name), which are then passed on to the OCU responsible for the area where the call originated. The Operations Centre within the OCU in question will then review the incident priority and allocate resources to respond to it. In the case of 'Immediate Response' incidents, the Police are required to attend the scene within 10 minutes.

Civilian services

Three command and control applications areas were analysed in the civilian services: air traffic control, the rail network and the national electricity grid.

Air Traffic Control

Air Traffic Control is a highly evolved process based on clearly defined procedures. The procedures used in normal operations are based on the aircraft flight plan, which describes its intended route. This route includes the starting location, beacons or reporting points that it will pass and its final destination airport. This information derives from the flight data strip computer and is presented to the controllers in the form of a flight progress strip by flight strip assistants. The flight data strip contains coded information showing particulars about the aircraft and its route. From this information the controller can determine the approximate time and position at which the aircraft will arrive in the sector. UK controlled airspace is divided into sectors, each of which is monitored by an air traffic control team. As an aircraft travels through these sectors, responsibility for controlling it transfers from one controller to another. Making sure that aircraft pass through this airspace and take off and land safely is the key responsibility of individual controllers.

Railway maintenance

Three scenarios were analysed in the UK rail industry. The activities under consideration were those involved in the setting up of safety systems required when carrying out maintenance of track. Under normal conditions a signaller has the key responsibility for controlling train movements and maintaining safety for an area of railway line. This control occurs remotely from the line at a control centre (a signalbox or signalling centre). These can be located many miles from where activity could be taking place. During maintenance, another person takes responsibility (possession) for an area of the line. Communication and coordination is required to transfer responsibility between the signaller and track maintenance engineers. The track maintenance engineers also have to communicate and coordinate with various other personnel, such as those carrying out maintenance within their areas of control, drivers of trains and on track-plant which may be in the zone where maintenance is taking place (called the possession), and personnel implementing aspects of the possession (all of which may also be dispersed over a certain geographical area). The three specific maintenance scenarios analysed were: planned maintenance (the processes and activities for setting up a possession for a stretch of track so that planned maintenance can take place), emergency engineering work (the processes and activities for unplanned emergency engineering work on the line, such as when track or infrastructure has been damaged or has suddenly degraded) and ending track possession (the reversal of the processes and activities for planned maintenance).

National Grid (high voltage electricity)

Three scenarios were analysed in the bulk electricity transmission industry (the National Grid). National Grid Transco own, maintain and operate the high voltage electricity transmission system in England and Wales. This complex and distributed system is comprised of 4500 miles of overhead lines, 410 miles of underground cables and 341 substations. The scenarios under analysis were observed at the Network Operations Centre (NOC) control room and in a number of geographically remote substations. Two outage scenarios were observed. There were three main parties involved in these, a party working at Substation A on the Substation B circuit, a party working at Substation B on the Substation A circuit (that is at either end of a 30 mile overhead line) and an overhead line party working in between. A return to service scenario was also observed. This scenario also involved a circuit between substations and the NOC. Observation focussed upon six main parties that were involved and the complex technological infrastructure that facilitated remote operations and communication.

Armed services

Command and control applications were analysed within all three armed services: army, air force and navy.

Navy

The Royal Navy allowed a team of researchers access to one of their training establishments – the Maritime Warfare School – at HMS Dryad in Southwick, Hampshire. Observations were made during Command Team Training (CTT). This programme involved training the Command Team of a warship in the skills that would be necessary for them to defend their ship in a multi-threat environment; 'assimilate, interpret and respond correctly to the information received from external sources while reporting, directing and managing their and other units in the joint conduct of maritime operations' (Hoyle, 2001, p.3). The training programme was conducted in a representative Type 23 Ship 'Operations Room Simulator' (ORS). The simulator room is slightly enlarged to allow for staff observation but otherwise was to scale. In addition to this simulator room there was a room which included a team of personnel who helped make any threats seem more realistic, that is they portrayed other ships and aircraft as well as personnel from other parts of the ship. Three scenarios (air threat, subsurface threat and surface threat) were observed. The Anti-Air Warfare Officer (AAWO) was the main agent observed in the air threat scenario and the PWO (Principal Warfare Officer) was the main agent observed for the subsurface and surface threat scenarios. Other agents were heard and seen interacting with either the AAWO or the PWO. The AAWO and PWO formed two central hubs in a split network architecture (as defined by Dekker, 2002), as shown in chapter 4.

Air Force

The scenario analysed in the RAF took place on board an E3D AWACS (Airborne Warning and Control System) aircraft and covers the operations for a simulated war exercise. The RAF ran a training course for Combined Qualified Weapons Instructors (CQWI). The evaluation of this training course took place over a two-week period. A simulated war exercise was carried out each day involving three key teams of personnel: ground based support, the E3D (AWACS) team and the fighter pilots. Ground based support includes all personnel assigned to the mission who are not flying in the simulated war exercise. The term 'fighter pilots' refers to both fighter and bomber pilots who took part in the exercise. These pilots flew a variety of aircraft, which numbered between 20 and 40 for each mission. The purpose of the E3D team was to provide support for both ground and fighter personnel. Their role involved providing a global picture of the war from the sky as it developed. This information was relayed to ground support staff and to individual fighter pilots. All personnel were kept up to date with where fighters were in relation to one another and of any fatalities. There were 18 crewmembers that operated the aircraft and made surveillance and support for ground and fighters possible. Observers were present on board the E3D aircraft and monitored the communications of a number of key people throughout the exercise. From this a distributed network architecture was observed (as defined by Dekker, 2002), as shown in chapter 5

Army

The study of command and control in the Army took place at the Command And Staff Training (CAST) exercises at the British Army's Land Warfare Centre in Warminster. Observations of both Brigade Headquarters and Battle Group Headquarters were undertaken. The studies were focused on the planning process, known as the Combat Estimate, war-gaming and simulation of the enactment of the plan. The plan is considered adequate when it meets the commander's intent, provides clear guidance to all sub-units and enough detail to allow the effects of the available combat power to be synchronised at critical points (MoD, 2005b). Flexibility is described in terms of the agility and versatility required to respond to the situation (and enemy) as events occur. Timeliness, finally, is about ensuring that there is 'sufficient' time for the battle procedure to be enacted. The Combat Estimate is summed up (and often referred to) as the seven questions. These questions break down the process by which plans are made and actions taken; they summarise the activities and outcomes of the different stages of the process. The structural relationships between the different members of the team revealed a centralised network with sharing (as defined by Dekker, 2002), as was show in chapter 6.

Common Features of the Domains and Application of Command and Control

Despite the differences in the domains, the command and control applications share many common features. That is at some level they can be regarded as 'generic'.

- First, they are typified by the presence of a central control room that is remote from the primary operations. Data from the field are sent to displays and/or paper records about the events as they unfold over time.
- Second, there is (currently) considerable reliance on the transmission of verbal messages between the field and the central control room. These messages are used to transmit reports and command instructions.
- Third, a good deal of the planning activities occur in the central control room, which are then transmitted to the field. There are collaborative discussions between the central control room and agents in the field on changes to the plan in light of particular circumstances found in-situ.
- Fourth, the activities tend to be a mixture of proactive command instructions and reactive control measures.
- Finally, different social architectures are readily supported, such as centralised, split and distributed network.

It is hypothesised that one of the determinants of the success or failure of a command and control system will be the degree to which both the remote control centre and agents in-the field can achieve shared situational understanding about factors such as: reports of events in the field, command intent, plans, risks, resource capability, and instructions. This places a heavy reliance on the effectiveness of the communications and media between the various parties.

Taxonomies of Command and Control Activities

Analysis of the task analyses from these three domains led to the development of a taxonomy of command and control activities, as indicated in Table 7.1. The resultant data from the observational studies and task analyses were subject to content analysis, in order to pick out clusters of activities. These clusters were subjected to thematic analysis consistent with a 'grounded theory' approach to data-driven research. It was possible to allocate most of the tasks in the task analysis to one of these categories. To this extent, the building of a generic model of command and control was driven by the data from the observations and task analyses.

Table 7.1   Taxonomy of command and control activities

Category	Table No.	Definition of activities
Receive	7.2	Receipt of data or information, a request or an order.
Plan	7.3	Planning activities and planning decisions.
Rehearse	7.4	Rehearsal of plan prior to action.
Communicate	7.5	Transfer of verbal, written or pictorial information.
Request	7.6	Request for data and information or assistance.
Monitor	7.7	Monitoring and recording of effects of plan implementation.
Review	7.8	Reviewing the effectiveness of plans or actions.

The detailed taxonomies may be found in the following seven tables. The 'receive' taxonomy, as shown in Table 7.2, identifies activities that are associated with receiving orders, requests, data and information that relate to past, present or future events. This information can act as a trigger for new command and control tasks, or modifications of ongoing tasks. Thus the information may be either feedforward or feedback.

The 'planning' taxonomy, as shown in Table 7.3, describes all of the activities associated with the preparation, assessment and choice of the plan. These activities include gathering of information, assessing options, discussing effects and prioritising alternative courses of action.

The 'rehearsal' taxonomy, as shown in Table 7.4, identifies activities that are associated with rehearsal of the plan prior to implementation. Most of the domains discuss the plan with the other parties, with the exception of ATC. The army also run a wargame on a map to consider the synchronisation of potential effects and likely courses of enemy responses.

The 'communicate' taxonomy, as shown in Table 7.5, refers to all of the activities associated with remote communication from the control centre. When the plan is communicated verbally, there is a read-out and read-back procedure, which may also act as a verbal rehearsal, although it does not formally belong in the 'rehearsal' taxonomy.

The 'request' taxonomy, as shown in Table 7.6, refers to the manner in which the command and control centre asks for information and support from other parties. This includes agents in the field, other agencies, and other personnel in the command centre.

The 'monitor' taxonomy, as shown in Table 7.7, refers to all of the activities associated with keeping track of the changing situation and events being performed remotely. These activities include recording any changes to the plan as they occur.

Table 7.2 The 'receive' activities taxonomy

Domains	Emergency Services		Civilian Services			Armed Services		
	Police	Fire	NATS	NGT	NR	Army	Navy	Air
Incoming calls	■	■	■	■	■	■	■	■
Paper message		■			■	■	■	■
Face to face			■	■		■	■	■
Diary of work				■				
Incoming alarms	■		■	■	■	■	■	■
Identity exchange	■		■	■	■	■	■	■
Live displays			■	■	■	■	■	
Pre planned/pre-defined activities	■		■	■	■			
Database			■	■				
Handover	■		■		■			
Procedures/systems (implicit comms)	■				■			

Table 7.3   The 'planning' activities taxonomy

Domains	Emergency Services		Civilian Services			Armed Services		
	Police	Fire	NATS	NGT	NR	Army	Navy	Air
Review of location of assets								
Establish status of assets and resources								
Request status of current activities								
Gather information (site, intel, environs)								
Integrate information								
Get people to site								
Develop mission timings								
Identify areas of interest								
Identify decision points								
Undertake environmental analysis								
Determine options (own and enemy)								
Identify potential conflicts								
Identify tactics (own and enemy)								
Check if information is sufficient								
Consult other parties								
Discuss effects								
Assess options								
Select between alternative plans								
Assign assets to tasks								
Assess risks with plan								
Communicate plan to other parties								

Development of a Generic Process Model of Command and Control

Table 7.4 The 'rehearsal' activities taxonomy

Domains	Emergency Services		Civilian Services			Armed Services		
	Police	Fire	NATS	NGT	NR	Army	Navy	Air
Discuss plan verbally	■			■	■	■	■	■
Move assets on map to rehearse plan						■		

Table 7.5 The 'communicate' activities taxonomy

Domains	Emergency Services		Civilian Services			Armed Services		
	Police	Fire	NATS	NGT	NR	Army	Navy	Air
Exchange identities	■	■	■	■	■		■	■
Issue instructions/orders	■	■	■	■	■	■	■	■
Read-back instructions	■	■	■	■	■	■		
Confirm read-back	■		■	■	■	■		
Record date and time of instructions	■	■	■	■	■		■	■

Table 7.6   The 'request' activities taxonomy

Domains	Emergency Services		Civilian Services			Armed Services		
	Police	Fire	NATS	NGT	NR	Army	Navy	Air
Request status from personnel	■	■		■	■	■	■	■
Request information from other parties	■	■	■	■			■	■
Pass information onto other parties	■	■	■			■	■	■
Request additional resources or assets	■	■	■		■	■	■	
Request support from other services	■	■	■			■	■	

Table 7.7   The 'monitor' activities taxonomy

Domains	Emergency Services		Civilian Services			Armed Services		
	Police	Fire	NATS	NGT	NR	Army	Navy	Air
Track assets (and enemy)	■	■	■	■	■	■	■	■
Identify conflicts with plan	■	■	■		■	■		■
Allocate resource and assets to tasks	■	■				■	■	■
Control resources and assets	■	■	■		■	■	■	■
Record changes to plan	■	■	■		■	■	■	■

Development of a Generic Process Model of Command and Control   231

The 'review' taxonomy, as shown in Table 7.8, refers to all of the activities associated with an after-action review of the successful and less successful aspects of the activities. This includes formal procedures, informal records, incident reports and accident tribunals. The armed forces tend to be very thorough in applying this analysis at the end of every engagement, whereas the civilian and emergency services tend to be more informal unless there is an accident or near miss.

Table 7.8   The 'review' activities taxonomy (The taxonomies are related to comprehensive HTAs of each individual scenario in question. These can be found separately in the individual EAST analysis reports that deal with each live scenario.)

Domains	Emergency Services		Civilian Services			Armed Services		
	Police	Fire	NATS	NGT	NR	Army	Navy	Air
Effectiveness of actions	■	■				■	■	■
Deviation from plan					■	■	■	■
Key decision points	■	■	■	■		■	■	■
Internal updates and reports	■	■	■			■	■	■
Loss of situation awareness			■			■	■	■
Lessons learnt						■	■	■

Construction of the Model

From the taxonomies, and an analysis of the previous command and control models, it was possible to develop a generic process model as shown in Figure 7.1. Construction of the model was driven by the data collected through observation from the different domains, and the subsequent thematic analysis and taxonomic development. In the tradition of grounded theory the generic command and control model was as a result of our observations, rather than an attempt to impose any preconceived ideas of command and control. This may account for many of the differences in the current model developed in the course of the current research and those that have come before it.

It is proposed that the command and control activities are triggered by events at the top of the figure, such as the receipt of orders or information. These provide a mission and a description related to the current situation and events in the field. The gap between the mission and the current situation lead the command system to determine the effects that will narrow that gap. This in turn requires the analysis of resources and constraints in the given situations. From these activities plans are developed, evaluated and selected. The chosen plans are then rehearsed before being communicated to agents in the field. As the plan is enacted, feedback from the field is sought to check that events are unfolding as expected. Changes to the mission or the events in the field may require the plan to be updated or revised. When the mission has achieved the required effects, the current set of command and control activities may come to an end.

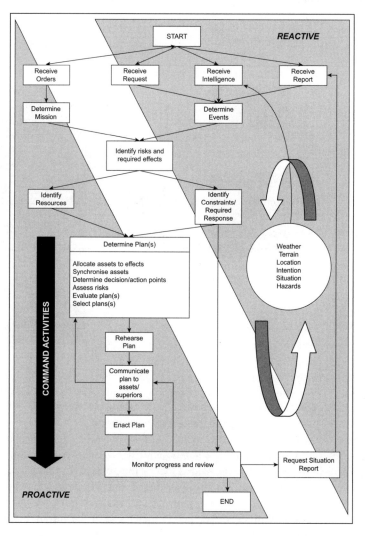

Figure 7.1 Generic process model of command and control

The model in Figure 7.1 distinguishes between 'command' activities, in the shaded triangle on the left-hand side of the figure, and 'control' activities, in the shaded triangle on the right-hand side of the figure. Command comprises proactive, mission-driven planning and co-ordination activities. Control comprises reactive, event-driven, monitoring and communication activities. The former implies the transfer of mission intent whereas the latter implies reaction to specific situations.

Network Enabled Capability

Network Enabled Capability invariably leads to a decentralisation of command, often right down to the level of personnel in the field. This suggests a fundamental shift in the way that command is performed. Command becomes much less rigid and prescriptive and rather more to do with setting goals and rules. In the field scenarios that have been examined such 'networks' are uncommon. The Fire Service (where the arriving commander who is closest to the incident will take charge of it) and perhaps to some extent Air Traffic Control (where individual controllers decide and negotiate amongst themselves) are two of the more cogent examples. However, in neither of these domains would one expect to see highly distributed communications – in ATC the primary communications remain between pilot and controller, and in the Fire Service the communications are either between Incident Commander and Sector Commanders, or between Incident Commander and control room. How does the process model take account of NEC trends such as these? Clearly within the model the critical difference between a 'decentralised' and 'centralised' paradigm is the level at which 'command' reaches from the top to the bottom. For example, under a centralised paradigm one can imagine 'command staff' identifying resources, whereas under a decentralised paradigm 'the person on the ground' may perform the same task. The key point appears to be that centralisation/decentralisation (that is NEC) does not appear to change the 'process' (model) merely the 'ownership' of the process elements. Indeed, the approach of mission command in the military implies a gradual focusing from strategic to operational definition of a particular objective: at a command node in the network, a strategic objective is defined in broad terms, for example 'secure a supply route between point A and point B' and this is passed down the hierarchy to nodes that cast the objective in tactical terms, for example in terms of assigning resources to the objective etc., and this is further passed down to the nodes that will perform the operations required to achieve the objective. One issue for NEC (and related concepts of highly distributed command) is where the strategic objective might arise if there is no central command. One response to this is that often objectives can be created opportunistically, for example a particular course of action might become possible due to unforeseen or unexpected circumstances and there might be only a limited time-window in which to perform the action, and seeking permission would delay the operation. To some extent, mission command provides officers at the operational level the capability to respond in opportunistic fashion. However, one would not wish for an entire mission to be conducted in such a manner and would expect that any decisions to operate off-mission be justifiable, in terms of Rules of Engagement, management of risk and appropriate use of resource. The question is whether our model is able to reflect such command activity.

Model Validity

As the process model is derived from the analysis of the three different domains, it should also explain the activities that go on in each of them.

In the military domain, such as the army, the scenario begins with orders from a higher level in the command chain. These orders form the mission that has to be turned into a plan. The plan starts as an outline of the required effects on a map. These effects have to be turned into alternative courses of action that would result in the required effects, taking any intelligence of what the enemy is doing into account. Each of the courses of action is evaluated in terms of the resources required and possible risks. The most optimal plan is selected, although alternative courses of action are kept in reserve. The plan is rehearsed though a wargaming exercise and then communicated to the assets in the field, who might have to undertake lower level planning to meet their mission. If all levels of the organisation are content with the plan then it is put into action. Regular field reports are requested and the effects of the changing situation are fed back to check if events are unfolding as anticipated. Deviations from planned events may require more planning to determine if changes in the course of action are required. Changes in the plan are fed up and down the command chain. When the mission objectives have been achieved, the scenario ends.

In the emergency services domain, such as the fire service, the scenario begins with an emergency call from a member of the public, or from another emergency service. As much information as possible about the event is recorded so that preliminary resource allocation may be undertaken. Template plans for different types of event can be applied, such as 'Domestic House Fire', 'Road Traffic Accident, and 'Chemical Tanker Spillage'. This can operate as a means of guiding the initial responses to the emergency, and getting the right kind of vehicles, equipment and people to the event. If the event turns out to be as expected, much of the incident planning can be left to the field operative, such as determining the appropriate course of action and implementing the plan. Feedback to central command will confirm this. If the event turns out to be more extensive or different from that anticipated, then details of the scenario may have to be fed back to central command and the plan worked out there first. Then the chosen course of action will be communicated to the assets in the field and regular reports requested to check that the effects are as anticipated. If a new situation comes to light, then new courses of action may have to be planned for. More complex incidents may require multi-agency cooperation, which will also be managed from central command. When all danger is removed from the situation and the incident is cleaned up, the scenario may be ended.

In the civilian domain, such as ATC, the scenario begins with receipt of a call from an aircraft to request clearance to enter specified airspace. The mission objectives are overarching for all scenarios and are expressed as the safe and efficient transport of aircraft within and between sectors. There are rules for aircraft separation and set procedures if this separation is compromised. An aircraft may request a descent so that it can land at an airport. The ATCO has to assess the effects and risks, and determine a sequence of actions that will achieve the desired outcome. The exact set of actions will be affected by a number of factors, such as the workload of the ATCO, weather conditions, and number of aircraft and complexity of airspace. The ATCO will, in effect, mentally anticipate the likely outcome of the planned actions for the aircraft in question and those in the near airspace. The chosen course of action will be written onto the flight strip and communicated to the pilot. The rules require a read-back for every instruction given. The ATCO will monitor the progress of the aircraft, to check that the pilot is progressing as planned. If the aircraft does not appear to be making progress, the ATCO will request a situation update from the pilot. The scenario ends when the aircraft leaves the sector being monitored by the ATCO.

Summary

Thus it is possible to use this model to:

- Provide a common platform for reviewing command and control activities in disparate domains.
- Explain some of the complexity of different command and control domains.
- Simplify some of that complexity.

Conclusions

The generic process model of command and control developed in the course of the research described in this book appears to have some differences to the C4i specific models that were highlighted in the review of modelling literature. These individual models were (amongst others) Lawson's (1981) control theoretic model, Hollnagel's (1993) control modes model, Rasmussen (1974) and Vicente's (1999) decision ladder model, and Smalley's (2003) functional command and control model.

The model in Figure 7.1 contains all of the information processing activities within Lawson's model (shown earlier in Figure 2.2, see chapter 2), but with greater fidelity and relevance to command and control activities. There is no explicit representation of a 'desired state' in the newer mode; rather this is expressed in terms of 'required effects', which may be open to change in light of changes in the mission or events in the world. It is also worth noting that the 'desired state' remains static in Lawson's model, which is a weakness of the approach.

Whilst much of the command and control activities are implicitly listed in Hollnagel's model (shown in Figure 2.24, see chapter 2) under 'goals', 'plans', 'execution' and 'events', the new model makes all of these activities explicit. The model does not indicate the effects of temporal change on the command and control activities. The model does however distinguish between the proactive 'command' activities and the reactive 'control' activities. It is probable that in higher tempo situations the command and control system is more likely to be in a reactive mode (that is the right-hand side of Figure 7.1). Conversely, it is probable that in lower tempo situations the command and control system is more likely to be in a proactive mode of operation (that is the left-hand side of Figure 7.1).

The new task based model does not attempt to distinguish between knowledge and system states in the way that Rasmussen and Vicente's decision ladder model does (see chapter 2). The decision ladder model has the inherent flexibility of shortcuts (which the analyst is required to identify), which can be used to indicate different levels of expertise. This makes the decision ladder model a better explanation of individual behaviour, but less applicable to the description of the activities in a command and control system. The generic decision ladder model could be used to describe any system, but it lacks the fidelity of Smalley's model for command and control. The decision ladder model does not attempt to distinguish between command and control activities, nor proactive and reactive behaviour.

As with Smalley's model (shown in Figure 2.20, see chapter 2), the new model does distinguish between 'command' activities and 'control' activities, and between 'internal' and 'external' co-ordination. The new model does not attempt to distinguish between 'information processing' and 'decision support' functions, primarily as the purpose of the generic model was to remain independent of technology and allocation of function. As with Smalley's model, the newer model also has a high degree of fidelity with regard to the command and control activities. This was seen as a strength of both of the models. Whereas Smalley's model was based solely on military command and control the newer model was based on a broader domain base. Both models provide a basis for research investigations into command and control activities.

Military Command and Control can be seen to progress through three distinct phases of activity, broadly; information retrieval, decision making, then dispersion and action. It differs from civilian examples of command and control (that have been studied previously) in terms of context. Firstly, the civilian operational environment and, therefore, to some extent the management infrastructure, tends to be relatively static despite the high tempo (and workload) often found; generally, it does not reconfigure itself quite so dramatically. This appears to arise from an emphasis on minimising system disturbances, and maximising safety, as opposed to military effects that are designed to yield disturbances in equilibrium. Secondly, of course, the biggest difference between civilian and military Command and Control is that there tend not to be opposing forces in civilian settings (albeit there is commercial competition), also conducting their own Command and Control activities that are actively trying to defeat, disrupt, evade and mislead. This is quite a big difference and probably explains its more 'dynamic' (and high tempo) nature of military activity. As a result of operational constraints and highly defined procedures, civilian command and control tasks tend to occur in a generally repeatable and often continuous manner. By contrast, in the military domain the tactical and strategic levels of control are often much closer (conceptually and physically) to the operational arena, and the feedback loops within the task network reveal the degree of flexibility and reconfigurability inherent in this arrangement. This is well suited to an equally dynamic and reconfiguring environment and to a much wider range of possible and available effects. These contextual differences could be why the civilian domain is thought to lead the way in NEC capabilities (air traffic control is one example where a distributed communications infrastructure and peer-to-peer command architecture is already in place). The success of such systems in the civilian domain could be simply due to the constraints that can be imposed on the context. It is certainly the case that agents and artefacts tend to behave in highly prescribed ways. It can be argued that technology is challenged to realise similar advantages in a dynamic and reconfigurable military context. The emergent properties revealed by the EAST method are designed to be helpful in this respect.

Within the EAST method the HTA, CDA and CUD combine with the social networks to show the external conditions that need to be responded to, the actors in the scenario who need to respond, the dimensions of teamwork that are applicable and the efficacy of the mediating technology. Using this combination of component methods it becomes possible to begin to specify how NEC approaches might practically be deployed to achieve the desired goal. The specification can be based on alternate configurations of the social network and the manner in which the underlying communications infrastructure may facilitate or enhance such a configuration. The propositional networks also show what knowledge needs to be shared and permits analysis of how well communications technology may (or may not) facilitate this. The current thinking situates the EAST method within a descriptive level of analysis, but the examples above serve to show how it can be used as a form of modelling tool. In the design and specification of NEC according to the principals of joint cognitive systems, it becomes possible to subject some of the outputs of the EAST method to known inputs, and to analyse their propagation and affect at the systems level. It is argued that that the results of these known inputs, at the systems level, 'emerge' from the complex interactions of systems level phenomena related to task, social and propositional networks. EAST provides a way of capturing them.

By conducting observations across several domains, the aim of this work has been to develop a generic framework for command and control. In order to progress this into a coherent theory the next phase of the work is to explore how the various domains perform operations within each heading and to ask how the removal or disruption of activity under a heading will impair performance within a given domain. An approach to this would be to employ the WESTT (workload, error, situational awareness, time and teamwork) tool (see McMaster et al., 2005).

Bibliography

Ainsworth, L. and Marshall, E. (1998). 'Issues of quality and practicality in task analysis: preliminary results from two surveys.' *Ergonomics* 41(11), 1604–1617. Reprinted in J. Annett and N.A. Stanton (2000) *Task Analysis*. London: Taylor and Francis, pp.79–89.

Alberts, D.S. and Hayes, R.E. (2006). *SAS-050, Exploring New Command and Control Concepts and Capabilities* <http://www.dodccrp.org/SAS/SAS-050%20Final%20Report.pdf>.

Annett, J. (2004). 'Hierarchical task analysis.' In N.A. Stanton, A. Hedge, K. Brookhuis, E. Salas, and H. Hendrick (eds) *Handbook of Human Factors methods*. London: Taylor and Francis.

Annett, J. and Stanton, N.A. (2000). *Task Analysis*. London: Taylor and Francis.

Arnold, J., Cooper, C.L. and Robertson, I.T. (1995). *Work Psychology: Understanding Human Behaviour in the Work Place* (2nd edition). London: Pitman.

Baber, C. and Stanton, N.A. (1996). 'Human error identification techniques applied to public technology: predictions compared with observed use.' *Applied Ergonomics* 27(2), 119–131.

Baber, C. and Stanton, N.A. (2004). 'Methodology for DTC-HFI WP1 field trials.' *Defence Technology Centre for Human Factors Integration*, Report 2.1.

Baber, C., Houghton, R.J., McMaster, R., Salmon, P., Stanton, N.A., Stewart, R.J., and Walker, G. (2004b). 'Field studies in the emergency services.' *HFI-DTC Technical Report/WP 1.1.1/1-1,* November 2004.

Baber, C., Walker, G., Stanton, N.A. and Salmon, P. (2004a). 'Report on initial trials of WP1.1 Methodology conducted at fire service training college.' *HFI-DTC Technical Report, WP1.1.1/01*, 29 January 2004.

Bar Yam, Y. (1997). *Dynamics of Complex Systems*. Jackson, TN: Perseus.

Bar Yam, Y. (1997). *Dynamics of Complex Systems*. Jackson, TN: Perseus cited in Grand, S. (2000). *Creation: Life and How to Make It*. Phoenix: London.

Bell, H.H. and Lyon, D.R. (2000). 'Using observer ratings to assess situation awareness.' In, M.R. Endsley (ed.) *Situation Awareness Analysis and Measurement* (pp.129–146). Mahwah, NJ: LEA.

Borgatti, S.P., Everett, M.G. and Freeman, L.C. (2002). *UCINET 6 for Windows*. Harvard: Analytic Technologies. A 30-day free trial version is available on the web at <http://www.analytictech.com>.

Builder, C.H., Bankes, S.C. and Nordin, R. (1999). *Command Concepts: A Theory Derived from the Practice of Command and Control*. Santa Monica, CA: Rand.

Burke, S.C. (2005). 'Team task analysis.' In, N.A. Stanton et al. (eds), *Handbook of Human Factors and Ergonomics Methods* (pp.56.1–56.8). London: CRC.

Burt, R.S. (1991). STRUCTURE. Version 4.2. New York: Columbia University.

Chin, M., Sanderson, P., Watson, M. (1999). 'Cognitive work analysis of the command and control work domain.' *Proceedings of the 1999 Command and Control Research and Technology Symposium* (CCRTS), June 29 – July 1, Newport, RI, USA, Volume 1, pp.233–248.

Choisser, R. and Shaw, J. (1993). 'Headquarters effectiveness assessment tool.' In C. Jones (ed.) *Toward a Science of Command, Control, and Communications: Progress in Astronautics and Aeronautics*, Vol. 156. Washington: Institute of Aeronautics and Astronautics, Inc.

Clancey, W.J. (1993). 'The knowledge level reinterpreted: Modelling socio-technical systems.' *International Journal of Intelligent Systems* 8(1), 33–50.

Dekker, A.H. (2002). *C4ISR Architectures, Social Network Analysis and the FINC Methodology: An Experiment in Military Organisational Structure*. DSTO Electronics and Surveillance Research Laboratory. DSTO-GD-0313.

Diaper, D. and Stanton, N.A. (2004). *Handbook of Task Analysis in Human-Computer Interaction*. Mahwah, NJ: Lawrence Erlbaum Associates.

Dockery, J.T. and Woodcock, A.E.R. (1993). 'No new mathematics! no new C2 theory!' In, C. Jones (ed.) *Toward a Science of Command, Control, and Communications: Progress in Astronautics and Aeronautics*, Vol. 156. Washington: Institute of Aeronautics and Astronautics, Inc.

Doebelin, E.O. (1972). *System Dynamics: Modelling and Response*. Columbus, Ohio: Charles E. Merrel Publishing Company.

Driskell, J.E. and Mullen, B. (2005). 'Social network analysis.' In N.A. Stanton et al. (eds) *Handbook of Human Factors and Ergonomics Methods* (pp.58.1–58.6). London: CRC.

Drury, C.G. (1990). 'Methods for direct observation of performance.' In, J.R. Wilson and E.N Corlett (eds) *Evaluation of Human Work – A practical ergonomics methodology* (pp.35–57). London: Taylor and Francis.

Duncan, W.J. (1981). *Organizational Behaviour, 2nd Edition*. Boston, Mass.: Houghton Mifflin.

Edmonds, B. and Moss, S. (2005). *The Importance of Representing Cognitive Processes in Multi-Agent Models* <http://bruce.edmonds.name/repcog/>.

Endsley, M.R. (1995). Toward a theory of situation awareness in dynamic systems. *Human Factors* 37(1), 32–64.

Flanagan, J.C. (1954). 'The critical incident technique.' *Psychological Bulletin* 51, 327–358.

General Dynamics (2003). *COMBAT use case model*. Oakdale, South Wales: General Dynamics UK Ltd.

Harris, C.J., and White, I. (1987). *Advanced in Command, Control and Communication Systems*. London: Peregrinus.

Hitchins, D.K. (2000). *Command and Control: The Management of Conflict* <http://www.hitchins.net/CandC.html>.

Hollnagel, E. (1993). *Human Reliability Analysis: Context and Control*. London: Academic Press.

Houghton, R.J., Baber, C., Cowton, M., Stanton, N.A. and Walker, G.H. (In Press). 'WESTT (Workload, Error, Situational Awareness, Time and Teamwork): An analytical prototyping system for command and control.' *Cognition, Technology and Work.*

Houghton, R.J., Baber, C., McMaster, R., Stanton, N.A., Salmon, P., Stewart, R.J., and Walker, G. (2006). 'Command and control in emergency services operations: A social network analysis.' *Ergonomics* 29(12–13*)*, 1204–1225.

Hoyle, G. (2001). *FCTT WP 10, 13 and 15 – Support Facilities.* Report No. SEA/00/ TR/2296 – CH575/FCTT/880028. SEA Ltd Internal Report.

Janis, I.L. (1982). 'Counteracting the adverse effects of concurrence-seeking in policy planning groups: theory and research perspectives.' In, H. Brandstatter, J.H. Davis and G. Stocker-Kreichgauer (eds) *Group Decision Making.* London: Academic Press.

Jones, C. (1993). *Toward a Science of Command, Control, and Communications. Progress in Astronautics and Aeronautics*, Vol. 156. Washington: Institute of Aeronautics and Astronautics, Inc.

Joslyn, C. (1999). *Semiotic Agent Models for Simulating Socio-Technical Organizations.* Los Alamos: Los Alamos National Laboratory.

Kaufman, A. (2004). *Curbing Innovation: How Command Technology Limits Network Centric Warfare.* Australia: Argos Press.

Kirwan, B. and Ainsworth, L.K. (1992). *A Guide to Task Analysis.* London, UK: Taylor and Francis.

Klein, G. and Armstrong, A.A. (2005). 'Critical decision method.' In, N.A. Stanton et al. (eds) *Handbook of Human Factors and Ergonomics Methods* (pp.35.1–35.8). London: CRC.

Klein, G.A., Calderwood, R., and MacGregor, D. (1989). 'Critical decision method for eliciting knowledge.' *IEEE Transactions on Systems, Man, and Cybernetics* 19(3), 462–472.

Kotter, J.P. (1978). *Organisational Dynamics: Diagnosis and Intervention.* Reading, Mass: Addison-Wesley.

Lawson, J.S. (1981). 'Command and control as a process.' *IEEE Control Systems Magazine*, March, 86–93.

Levine, D.K. (2005). *Economic and Game Theory: What is Game Theory?* <http://levine.sscnet.ucla.edu/general/whatis.htm>.

Marshall, A., Stanton, N., Young, M., Salmon, P., Harris, D., Demagalski, J., Waldmann, T., Dekker, S. (2003). *Development of the Human Error Template – a New Methodology for Assessing Design induced Errors on Aircraft Flight Decks.* Final Report of the ERRORPRED project (E!1970) to the UK Department of Trade and Industry, August 2003.

McMaster, R., Baber, C., and Houghton, R.J. (2005). *Investigating Alternative Network Structures for Operational Command and Control.* 10th International Command and Control Research and Technology Symposium, McLean Virginia, USA, June 2005.

Militello, L.G. and Hutton, J.B. (2000). 'Applied Cognitive Task Analysis (ACTA): A practitioner's toolkit for understanding cognitive task demands.' In, J. Annett and N.A. Stanton (eds) *Task Analysis* (pp.90–113). London, UK: Taylor and Francis.

Mill, J.S. (1843). *A System of Logic, Ratiocinative and Inductive.* London: John W. Parker.

Ministry of Defence (2000). *Human Factors Integration: An Introductory Guide.* London: HMSO.

Ministry of Defence (2004). *Network Enabled Capability: An Introduction.* London: HMSO.

Ministry of Defence (2005a). *Network Enabled Capability* (JSP 777). London: HM Stationary Office.

Ministry of Defence (2005b). *The Combat Estimate.* Warminster: MoD.

Ministry of Defence (Aug 2/3, 2005c). *Presentation 'Command and Staff Trainer (South) 1 WFR MiniCAST'.* Warminster: Land Warfare Centre.

Ministry of Defence (Aug 3, 2005d). *Presentation 'Wargaming: Mastering Your Enemy' (based on the Army Field Manual Vol. I (Combined Arms Operations) Part 2 (Jul 98) and 3(UK) Div Wargaming Aide Memoir)'.* Warminster: Land Warfare Centre.

Moray, N. (2004). 'Ou' sont les neiges d'antan?' In D.A. Vincenzi, M. Mouloua, and P.A. Hancock (eds) *Human Performance, Situation Awareness and Automation; Current Research and Trends.* Mahwah, NJ: LEA.

Morgan, G. (1986). *Images of organization.* London: Sage.

NATO (1988). *Glossary of Terms and Definitions* (STANAG AAP-6 (R)), Brussels, Belgium: North Atlantic Treaty Organisation.

Newell, A. and Simon, H. (1972). *Human Problem Solving.* Englewood Cliffs, NJ: Prentice-Hall.

Ogden, G.C. (1987). 'Concept, knowledge and thought.' *Annual Review of Psychology* 38, 203–227.

O'Hare, D., Wiggins, M., Williams, A. and Wong, W. (2000). 'Cognitive task analysis for decision centred design and training.' In, J. Annett and N.A. Stanton (eds) *Task Analysis* (pp.170–190). London: Taylor and Francis.

Pew, R.W. and Mavor, A.S. (eds) (1998). *Modeling Human and Organizational Behavior: Application to Military Simulations.* Washington, DC: National Academy Press.

Polson, P.G., Lewis, C., Rieman, J. and Wharton, C. (1992). 'Cognitive walkthroughs: a method for theory based evaluation of user interfaces.' *International Journal of Man-Machine Studies* 36, 741–773.

Rasmussen, J. (1974). *The Human Data Processor as a System Component: Bits and Pieces of a Model* (Report No. Risø-M-1722). Roskilde, Denmark: Danish Atomic Energy Commission.

Rasmussen, J. (1981). 'Models of mental strategies in process plant diagnosis.' In, J. Rasmussen and W.B. Rouse (eds) *Human Detection and Diagnosis of System Failures.* New York: Plenum Press.

Rasmussen, J. (1986). *Information Processing and Human Machine Interaction: An Approach to Cognitive Engineering.* New York, NY: North-Holland.

Rasmussen, J., Petjersen, A.M. and Goodstein, L.P. (1994). *Cognitive Systems Engineering.* New York: Wiley.

Reber, A.S. (1995). *Dictionary of Psychology.* London: Penguin.

Rescorla, R.A., and Wagner, A.R. (1972). 'A theory of pavlovian conditioning: variations in the effectiveness of reinforcement and no reinforcement.' In, A.H. Black and W.F. Prokasy (eds), *Classical conditioning II: Current research and theory* (pp.64–99). New York: Appleton-Century-Crofts.

Reynolds, C.W. (1987). 'Flocks, herds, and schools: A distributed behavioral model.' In, *Computer Graphics* 21(4) (SIGGRAPH '87 Conference Proceedings) pp.25–34.

Ross, D. (2005). *Economic Theory and Cognitive Science: Microexplanation.* Cambridge, MA: MIT Press.

Salmon, P.M., Stanton, N.A. and Walker, G. (2004b). 'National Grid Transco: Switching operations report.' *Defence Technology Centre for Human Factors Integration Report.*

Salmon, P.M., Stanton, N.A. and Walker, G. (2004c). 'National Grid Transco: Return to service report.' *Defence Technology Centre for Human Factors Integration Report.*

Salmon, P.M., Stanton, N.A., Walker, G., and Green, D. (2004a). 'Human factors design and evaluation methods review.' *Defence Technology Centre for Human Factors Integration*, Report No. HFIDTC/WP1.3.2/1.

Sanderson, P.M. (2003). 'Cognitive Work Analysis.' In J. Carroll (ed.) *HCI Models, Theories, and Frameworks: Toward an Interdisciplinary Science.* New York: Morgan-Kaufmann.

Shah A.P. and Pritchett A.R. (2005). 'Work environment analysis: Environment centric multi-agent simulation for design of socio-technical systems.' In P. Davidsson, et al. (eds) *Lecture Notes in Computer Science* 3415 (pp.65–77). Berlin: Springer Verlag.

Shepherd, A. (2001). *Hierarchical Task Analysis.* London: Taylor and Francis.

Smalley, J. (2003), 'Cognitive factors in the analysis, design and assessment of command and control systems.' In E. Hollnagel (Ed.) *Handbook of Cognitive Task Design* (pp.223–253). Mahwah, NJ: Lawrence Erlbaum Associates.

Stanton, N.A. (2004a). 'The psychology of task analysis today.' In D. Diaper and N.A. Stanton. (eds) *Handbook of Task Analysis in Human-Computer Interaction* (pp.567–584). Mahwah, NJ: Lawrence Erlbaum Associates.

Stanton, N.A. (2004b). 'Hierarchical task analysis: developments, applications and extensions.' *HFI-DTC Technical Report.*

Stanton, N.A. (2006). 'Hierarchical task analysis: developments, applications and extensions.' *Applied Ergonomics* 37, 55–79; invited paper to special issue on 'Fundamental Reviews of Ergonomics'.

Stanton, N.A. and Stevenage, S.V. (1998). 'Learning to predict human error: issues of acceptability, reliability and validity.' *Ergonomics* 41(11), 1737–1756.

Stanton, N.A, and Young, M.S. (1999). *A guide to methodology in ergonomics: Designing for human use.* London: Taylor and Francis.

Stanton, N.A., Ashleigh, M.J., Roberts, A.D. and Zu, F. (2001). 'Testing Hollnagel's contextual control model: Assessing team behaviour in a human supervisory control task.' *Journal of Cognitive Ergonomics* 5(1), 21–33.

Stanton, N.A., Salmon, P., Walker, G.H, Baber, C. and Jenkins, D. (2005). *Human Factors Methods: A Practical Guide for Engineering and Design.* Aldershot: Ashgate.

Stanton, N.A., Stewart, R.J., Harris, D., Houghton, R.J., Baber, C., McMaster, R., Salmon, P., Hoyle, G., Walker, G., Young, M.S., and Dymott, R. (2006). 'Distributed situational awareness in dynamic systems: Theoretical development and application of an ergonomics methodology.' *Ergonomics* 49(12), 1288–1311.

Stewart, R.J., Harris, D. Stanton, N.A., Salmon, P. Linsell, M. and Dymott, R. (2008). 'HMS Dryad: Air, surface and subsurface scenario report.' *Defence Technology Centre for Human Factors Integration Report.*

Stewart, R.J., Stanton, N.A., Harris, D., Baber, C., Salmon, P., Mock, M., Tatlock, K., Wells, L. and Kay, A. (2008). 'Distributed situational awareness in an airborne warning and control aircraft: application of a novel ergonomics methodology.' Special issue of *Cognition Technology and Work on Human Factors Integration* (in press).

Vicente, K.J. (1999). *Cognitive Work Analysis: Toward Safe, Productive, and Healthy Computer-Based Work.* Mahwah, NJ: Lawrence Erlbaum Associates.

Wainwright, J. and Mulligan M. (eds) (2004). *Environmental Modelling: Finding Simplicity in Complexity.* London: John Wiley and Sons Ltd.

Walker, G., Stanton, N.A and Young, M. (2002). 'Hierarchical Task Analysis of Driving (HTAoD).' In M.A. Hanson (ed.) *Contemporary Ergonomics.* London: Taylor and Francis.

Walker, G., Stanton, N.A., Wells, L., and Gibson, H. (2005). 'EAST methodology for air traffic control.' *Defence Technology Centre for Human Factors Integration Report* <http://www.hfidtc.com>.

Walker, G.H., Stanton, N.A., Gibson, H., Baber, C., Young, M. and Green. D. (2006a). 'Analysing the role of communications technology in C4i scenarios: A distributed cognition approach.' *Journal of Intelligent Systems* 15(1–4), 299–328.

Walker, G.H., Gibson, H., Stanton, N.A., Baber, C., Salmon, P. and Green, D. (2006b). 'Event analysis of systemic teamwork (EAST): A novel integration of ergonomics methods to analyse C4i activity.' *Ergonomics* 49(12–13), 1345–1369.

Watts, L.A., and Monk, A.F. (2000). 'Reasoning about tasks, activities and technology to support collaboration.' In, J. Annett and N. Stanton (eds) *Task Analysis* (pp.55–78). London: Taylor and Francis.

Wikipedia (2005). *Socio technical systems* <http://en.wikipedia.org/wiki/Socio-technical_systems>.

Index

abstraction decomposition space 39–40
activity analysis 39–41
agent models 32–6, 45–6, 48
air force 224 *see also* Royal Air Force (RAF) Boeing E3D Sentry case study
air traffic control 50, 222
army 50, 224 *see also* Battle Group HQ case study

Battle Group HQ case study
  activity stereotypes 200
  background 182–91
  co-ordination demands analysis (CDA) 192–6
  combat estimate 182, 183–9
  command and control in the British Army 181–2
  comms media 198
  comms usage diagrams (CUD) 195–9
  conclusions 219–20
  critical decision method (CDM) 209–19
  data sources 182
  Hierarchical Task Analysis (HTA) 192
  Network Enabled Capability (NEC) 197–9
  network links 204
  operation sequence diagrams (OSD) 204–10
  operational graphics 189–90, 198
  operational plans 191
  propositional networks 209–19
  radio communication 198
  seven questions 183–9
  social network analysis (SNA) 199–204
  social network metrics 200–203
  verbal communications 197, 198
  wargaming 190–91
battle winning ideas 186
Battlefield Area Evaluation (BAE) 183–5
bird flocking behaviour 34–5
Boids 34

case studies
  Battle Group HQ 181–220
  HMS Dryad 119–55
  Royal Air Force (RAF) Boeing E3D Sentry 157–80
Catastrophe Theory 14
Category Theory 14
CDA *see* co-ordination demands analysis
CDM *see* critical decision method
Cellular Automata 14
Chaos Theory 14
C4i (command, control, communication, computers and intelligence), definition 13
co-ordination demands analysis (CDA)
  advantages 89–90
  application times 91
  applications 87
  Battle Group HQ case study 192–6
  description 87
  disadvantages 90
  domain of application 88
  energy distribution 58–9
  example 90–91
  HMS Dryad case study 125–30
  procedure 88–9
  related methods 91
  reliability 91
  Royal Air Force (RAF) Boeing E3D Sentry case study 160, 162, 164, 178
  summary flowchart 92
  tools 91
  training 91
  use of 51
  validity 91
Cognitive Work Analysis (CWA) 39–43
command
  meaning of 10
  structures 26–8
command and control
  definition 10–11
  domains 221–5
  emergency services 221–2
  generic process model

conclusions 235–7
construction 231–3
domains 221–5
network enabled capability (NEC) 231–3
summary 235
taxonomies 225–31
validity of model 234–5
generic properties 11
structural models 13–19
taxonomies 225–31
comms usage diagrams (CUD) 51
advantages 93–4
application times 95–6
applications 92
Battle Group HQ case study 195–9
description 92
disadvantages 94
domain of application 92
energy distribution 61
example 94
procedure 92–3
related methods 94–5
reliability and validity 96
summary flowchart 96
tools 96
training 95–6
complexity, simplification of 8–9
constraints of models 9
contextual control model 43–5
Course of Action Development 186–8
critical decision method (CDM) 54
advantages 110
analysis example, energy distribution 62–7
application times 111
applications 108
Battle Group HQ case study 209–19
description 108
disadvantages 110
HMS Dryad case study 138
procedure 108–10
related methods 110
reliability 111
Royal Air Force (RAF) Boeing E3D Sentry case study 171
summary flowchart 112
tools 111
training 111
validity 111
CUD *see* Comms Usage Diagrams

cybernetic models 45, 48
cybernetic paradigm 13–19

dynamic models 25–32

EAST *see* event analysis of systemic team-work (EAST) methodology
emergence 33
emergency services 221–2
emergent properties of models 46
energy distribution 223
analysis example 57–66
application time 101–2
co-ordination demands analysis (CDA) 90–91
comms usage diagrams (CUD) 94
operation sequence diagrams (OSD) 106–7
propositional networks 115
related methods 101
reliability and validity 102
scenarios 50
social network analysis (SNA) 99–101
summary flowchart 103
tools 102
training 101
Entropy Computations 14
event analysis of systemic team-work (EAST) methodology
advantages 56–7
application time 67
co-ordination demands analysis (CDA) 51, 58–9, 87–92
comms usage diagrams (CUD) 92–6
component methods 51, 53
critical decision method (CDM) 108–12
disadvantages 57
domain of application 54
domains 50
example 57–66
hierarchical task analysis (HTA) 51, 79–86
method review summary 52–3
observation 51, 71–8
operation sequence diagrams (OSD) 103–8
procedure 54–6
propositional networks 112–17
reliability 67
review criteria 49, 51
scenarios 50

social network analysis (SNA) 96–102
software packages 70
summary of methodology 117–18
tools 70
training 66–7
validity of data 67, 70

FINC (Force, Intelligence, Networking and C2) methodology 25
fire service 50, 75–6, 221–2
fractals 14
Fuzzy Differential Equations 14
fuzzy sets 14

games theory 14

Headquarters Effectiveness Assessment Tool (HEAT) 16, 32
Hierarchical Task Analysis (HTA)
  advantages 80–81
  application times 85
  applications 79
  Battle Group HQ case study 192
  description of 20, 79
  disadvantages 81
  domain of application 79
  example 81–4
  HMS Dryad case study 124–5
  procedure 79–80
  related methods 84–5
  reliability 85
  summary flowchart 86
  tools 85
  training 85
  use of 51
  validity 85
Hitchens' N-squared (N^2) chart 19–20
HMS Dryad case study
  air threat CDA 126–7
  approach 124
  co-ordination demands analysis (CDA) 125–30
  command structure 122–3
  communications 123–4
  conclusions 154–5
  critical decision method (CDM) 138
  Hierarchical Task Analysis (HTA) 124–5
  operation sequence diagrams (OSD) 132–8
  overview 119–21
  propositional networks 138–54

scenarios 122
social network analysis (SNA) 128, 132
subsurface CDA 126–7
surface CDA 128
HTA *see* Hierarchical Task Analysis

Intelligence Preparation of the Battlefield (IPB) 183–5

Langrangian Formulation 14
Lawson's model 15

mathematical metrics 14
military aviation, scenarios 50
Mission Analysis 185
models 7–10
  abstracting reality 8
  agent 32–6, 45–6, 48
  complexity, simplification of 8–9
  constraints 9
  cybernetic 13–19, 45, 48
  dynamic 25–32
  emergent properties 46
  network 19–25, 45, 48
  results, synthesis of 10
  socio-technical 36–45, 46, 48
  structural 13–19

N-squared (N^2) chart 19–20
National Grid Transco (NGT) 57–66, 223
navy 50, 81–4, 223–4 *see also* HMS Dryad case study
Network Centric Operations Conceptual Framework (NCO CF) 17, 18
Network Enabled Capability (NEC) 12–13, 233–4
  Battle Group HQ case study 197–9
  benefits chain 46–8
network models 19–25, 45, 48

observation
  advantages 74
  application times 77
  disadvantages 74–5
  example 75–6
  related methods 77
  reliability 77
  summary flowchart 78
  tools 77
  training 77
  use of 51

validity 77
OODA Loop 16
operation sequence diagrams (OSD)
   advantages 105
   application times 107
   applications 103
   Battle Group HQ case study 204–10
   description 51, 54, 103
   disadvantages 105
   domain of application 103
   energy distribution 61–2
   example 106–7
   HMS Dryad case study 132–8
   procedure 103–5
   related methods 107
   reliability 107
   Royal Air Force (RAF) Boeing E3D Sentry case study 170–71, 179
   summary flowchart 108
   tools 107
   training 107
   validity 107
opportunistic control 43
organisation, definition 19
organisational charts 19
OSD *see* operation sequence diagrams

Partial Differential Equations 14
Petri Nets 14
police 50, 222
process charts 36–7
propositional networks
   advantages 115
   application times 116
   applications 112
   Battle Group HQ case study 209–19
   description 22–4, 54, 112
   disadvantages 115
   energy distribution 64–5, 68–70
   example 115
   HMS Dryad case study 138–54
   procedure 112–14
   related methods 115–16
   reliability 116
   Royal Air Force (RAF) Boeing E3D Sentry case study 172–7, 179–80
   summary flowchart 117
   tools 116
   training 116
   validity 116
pyramid structure metrics 21

radio communication 198
rail signalling 50
railway maintenance 222–3
Rasmussen's decision-ladder 41
Relativistic Information Theory 14
Royal Air Force (RAF) Boeing E3D Sentry case study
   approach 159–60
   co-ordination demands analysis (CDA) 160, 162, 164, 178
   command structure 158
   communications 158
   conclusions 177–80
   critical decision method (CDM) 171
   operation sequence diagrams (OSD) 170–71, 179
   overview 157
   propositional networks 172–7, 179–80
   scenario 159
   social network analysis (SNA) 160, 163, 166–70, 178–9
   task model 160, 161

SA (situational awareness) 22–3
SAS-050 Space of C2 11–12, 29–30
scrambled control 43
SCUD Hunt 25–8
semiotic agents 33–5
situational awareness (SA) 22–3
Skill, Rule, Knowledge (SRK) framework 43
Smalley's functional command and control model 37–9
social network analysis (SNA) 51
   advantages 99
   applications 96
   Battle Group HQ case study 199–204
   description 96
   disadvantages 99
   energy distribution 60
   example 99–101
   HMS Dryad case study 128, 132
   procedure 97–9
   Royal Air Force (RAF) Boeing E3D Sentry case study 160, 163, 166–70, 178–9
social networks 21–2
social organisation analysis 42–3
socio-technical models 36–45, 46, 48
SRK (Skill, Rule, Knowledge) framework 43

Stochastic Differential Equations 14
strategic control 43
strategies analysis 42
structural models 13–19, 17–19
System Dynamics 14

tactical control 43
task analysis 20
taxonomies of command and control 225–31
Threat Evaluation 183–5

verbal communications 197, 198

wargaming 190–91
WEA (Work Environment Analysis) 32
WESTT methodology 66, 67, 70, 105, 107, 237
work domain analysis 39–40
Work Environment Analysis (WEA) 32
worker competencies analysis 43

XML (extensible mark-up language) 32